Rend the Veil …
the Stage is Set

Bill McGreehan

KJV, Word Meanings, and Use of BCE (Before the Christian Era) and CE (Christian Era)

a. Unless otherwise designated, Scripture quotes are taken from the King James Version (KJV) and not the original 1611 King James English which sounds excessively archaic.
b. Word meanings provided are based on original Hebrew and Greek texts.
c. Unless indicated otherwise, any resemblance of names or characters to persons living or deceased is purely coincidental.
d. Supposedly for religious neutrality, since the 20th century secular publications have been increasingly adopting the 17th century scientific convention BCE (Before Common Era) and CE (Common Era) as alternatives to what had been common with BC (Before Christ) and AD (Anno Domini, or Year of Our Lord)—In my opinion, both these conventions are arbitrary and designed to satisfy no one. For the sake of bridging that divide, this book will abbreviate BCE to represent "Before the Christian Era" and CE for "Christian Era." For simplicity, the term CE is only employed for dates between the years 1 and 1000.

And, behold, the veil of the temple was rent in twain
from the top to the bottom;
and the earth did quake,...

Matthew 27:51

Dedication

In 1976 a group of believers met for prayer and Bible study at the home of Charles and Vicki Lake in Greenwood, Indiana. That marked the beginning of what is now Community Church of Greenwood (CCG). Recently I attended the Jubilee Celebration where Dr. Charles Lake (now retired) spoke to commemorate the start of CCG fifty years ago.

We joined CCG in 1978 during its infancy while meeting at a storefront building in downtown Greenwood. That period resulted in explosive growth of the church and our personal development in large part due to the Discipleship Training Program authored by Dr. Lake. Since that time Debby and I have made discipleship a priority.

As discussed in my prior book *Rend the Heavens . . . and Come Down* (2025), manna in the wilderness forms a metaphor of daily Bible reading, personal application, and prayer. Manna (Hebrew meaning - What is it?) had no precedent. By its' nature, manna necessitated the Children of Israel rise before the sun and gather it, that they might live in the wilderness. Bread from heaven contained everything they needed for the day. It sustained them forty years in the desert.

When someone asks for a definition of discipleship, my reply is, "A disciple makes disciples who make disciples…" Being a better Christian is not discipleship. Nor does it mean gaining followers. Disciples follow the command of Jesus to, "Do this in remembrance of Me."; emphasis on doing rather than just remembering.

Dr. Charles Lake continues to influence lives through discipleship work impacting Haitian believers. He certainly lives up to the legacy of making disciples who make disciples…

Acknowledgements

Stories and teaching material for *Rend the Veil . . . the Stage is Set* represent a sequel to the prior book *Rend the Heavens . . . and Come Down* (RtH). When friends ask Debby if she read the book RtH, she replies, "I don't need to read it; I lived it."

Both books reflect more than biographical statements; much of the content covers Biblical teaching derived from discipleship training and Bible study classes the past fifty years. When teaching a class, Debby always interjected humorous stories to lighten the mood. Like a team of sports commentators, I did the play-by-play and Debby added the color.

Personal stories and historical narratives form the framework for clarifying how current events set the stage for God's production of the soon return of Jesus Christ. *Rend the Heavens* and now *Rend the Veil* both focus on the necessity of resistance as a fitness center for believers.

Facing life-threatening trouble, Jesus questioned His disciples, "Why are ye fearful, O ye of little faith?" (Matthew 8:26) Personal challenges and events on the world stage need to be faced head-on and overcome by forcing the problem to become part of the solution.

My indebtedness to Chris Taylor (www.seraphimchris.com) for the outstanding book cover design. Also, artist Richard Eijkenbroek (www.Xquissive.ai) made a major contribution to *Rend the Veil.* Richard's artistic work created most of the illustrations based on my rough ideas.

All the World's a Stage

Contents

The Wrong Chapter

Introduction

No sooner had Mrs. Fairman begun mid-term testing in her English Literature class, but a sudden wave of dread gripped me. Upon glancing over the examination, I noticed the test covered a different chapter than I had studied the past few days.

For a brief moment, the implications of my mistake began to sink in. The past two years at Triangle Area High School (TAHS), I managed to improve a poor academic record into straight-A report cards every grading period.

As detailed in my recent book, *Rend the Heavens…and Come Down* (RtH, 2025), a series of miraculous events and dedicated study placed me academically near the top of the 1966 senior class. In the middle of my junior year, Mrs. Fairman related how she had recommended and been approved to have me moved mid-year from the lower to upper academic class based on three consecutive straight-A report cards. That occurred over the objection of administrators who claimed such a move had never before been done in the school district.

Now, after scraping the bottom academically for most of the high school years, my study habits and test results had improved dramatically. But suddenly I faced the prospect of failing a key English Literature exam due a simple oversight.

Only a Dream

Then, as quickly as it began, the classroom darkened as if the lights switched off. In the blackness a light and shadow flickered across the far wall. Gradually the image expanded to the size and shape of the east window and began to glide across the wall. Light from a car's headlight beams moved from the west wall over to the north wall, finally illuminating the flintlock and Kentucky long rifle replicas mounted above my bedroom desk.

Suddenly the realization hit that I had been dreaming. As headlights of the passing car rapidly exited the room, relief calmed the anxiety that gripped me. From time to time, similar dreams had disturbed my sleep, likely the result of the pressure to avoid falling back into scholastic failure. The Reverand Billy Graham once likened not being prepared to climbing the ladder of success, only to find it is against the wrong building.

That message sums up the theme of this book, with personal stories to add some context. A theatrical turn of events at the end of the book twists the plot on its' head as well as providing a foreshadowing of events to come at the return of Jesus Christ.

Village of Sidman

During the TAHS senior year in Sidman, Pennsylvania, as I prepared to apply for college and scholarships, Mrs. Fairman counseled me to broaden my interests beyond mathematics and science. Her advice met with lukewarm acceptance. Though I enjoyed art and writing, it did not go beyond subjects of space and rocket propulsion. Certainly, it did not extend to the school newspaper and yearbook, for which she was faculty adviser.

When Mrs. Fairman persisted, I agreed to join both teams while protesting that creative writing was outside my comfort zone. Rather than accepting my objection she suggested I learn to find topics and build stories around them. That concept served me well during the years ahead while employed developing jet and rocket engines.

Technical and Creative Writing

Throughout a fifty-year engineering career in gas turbine, hypersonic, and rocket propulsion, I experienced numerous opportunities to publish technical reports, manuals, and documentation as part of job responsibilities. Some of those publications involved thousands of pages of specifications for design, installation, and servicing of propulsion systems.

At the same time, doors opened to develop biblical and general interest teaching material on a wide range of subjects. Biblical training and discipleship topics were used in a weekly Bible Study at work as well as training classes and Sunday School lesson at several churches Debby and I helped to start.

Friends and colleagues suggested publishing some of the topics in a book. After a third and final professional retirement at the end of 2024, much of the teaching material and personal stories were published in the book *Rend the Heavens …and Come Down* (RtH). The format of RtH followed the framework of a road trip and cultural roadmap patterned after the 1971 song *American Pie* by Don McLean, but with a different outcome.

A recurring theme of the book RtH focused on the promised return of Jesus Christ at the opening of the sixth seal in Revelation 6:12-17. Throughout New and Old Testaments, Scripture employs stories and metaphors to portray spiritual principles. *Rend the Heavens* ended with a portrayal of the catching away of believers using a metaphor represented by a closing theater scene in the divine timetable for the church and history.

This sequel to *Rend the Heavens* expands the concept in the allegory of a theater production. The story unfolds the script of God's Word clarifying His plan for the Church and Kingdom through the ages. All of the stories in the following chapters lay out the plot, develop the characters, and describe how God has set the stage. Ultimately, God Himself directs the events according to His revealed Word.

As with the plot of any engaging story, the Biblical narrative includes many pivots and seeming contradictions. At the conclusion of His Script, God reveals how everything aligns with His stated purpose. Acts and scenes lead up to a climax of history as His plan is fulfilled.

Near the close of Daniel's prophetic revelation, God promised to rend open the veil signifying a crucial step to the completion of the age.

> And He said, Go thy way, Daniel: for the words are closed up and sealed until the time of the end. (Daniel 12:9)
>
> As God recorded by the prophet Isaiah,
>
> And He will destroy in this mountain the face of the covering cast over all people, and the vail [veil] that is spread over all nations. (Isaiah 25:7)

Instrumental to every good plot there appears an antagonist, and God's gospel message is no exception. Woven into His script, God reveals an adversary with a sinister purpose: the ultimate court jester. Characters in a narrative must be developed when injected into the plot. This story illustrates how God's adversary, though created perfect (Ezekiel 28:14-16), fell from his position through self-promotion to become the universal opponent of God and man.

The Director's Chair

The literal meaning of the Hebrew word *satan* is adversary or accuser. The name Satan in all languages confers his role as opponent of

God and His purpose. Characterization of God's adversary as the jester is not intended to trivialize nor to promote his position. Archangel Heylel's original name (Isaiah 14:12) derives from the Hebrew word *halal* signifying to praise or act foolishly.

Heylel also means morning star or to shine. Most English translations of the Bible employ the Latin name Lucifer which has the identical meaning as the Hebrew name Heylel. The Arabic equivalent is the name Zohran meaning morning star (pay attention here).

He fell from his position of archangel and leader of worship in heaven as a result of self-exaltation and pride (Isaiah 14:13-15). In his fallen position, Satan takes on the role of adversary to man and God, in resistance to God's eternal purpose. When playing a game or any competition, there must be an opponent. Even in the card game Solitaire, the odds serve as challenger.

Upon the final curtain, a stage is set, demonstrating the universal theme of the triumph of good over evil, light over darkness, truth over error. Christ's crucification and resurrection displayed God's full authority over His plan despite the opposition of Satan.

At the time of Jesus's sacrifice, it appeared to His disciples that the Messiah had succumbed to Satan, Jewish leaders, and the authority of Rome. On Resurrection Day, the first day of the week, Jesus triumphed and everything changed. Objects and events portend much more than their names imply: *aka* (also known as).

The rest of the gospel story serves only as the epilogue to the conflict. So also, the Book of Revelation unfolds God's power, just as Satan appears ready to snatch the crown from Jesus. In the sovereignty of God, each person will win or lose, depending on which side they stand. As Jesus declared,

> Enter ye in at the strait gate: for wide is the gate, and broad is the way, that leadeth to destruction, and many there be which go in thereat: Because strait is the gate, and narrow is the way, which leadeth unto life, and few there be that find it. (Matthew 7:13, 14)

The Narrow and Wide Gate

Campfire Story in the Mountains

Chapter One - Keep Your Powder Dry

DURING A CHURCH scouting campout a few years ago, I gathered our group of scouts around a campfire in the early evening to share a story. Sunlight cast shadows across a range of snow-covered mountain peaks in the distance, bordered by majestic pines and an alpine lake. Because the camping area stood adjacent to grizzly bear country, I arranged for three of the church scout leaders in their frontiersman outfits to stand guard around the camp armed with black powder rifles.

As the boys listened, I related the story of Jedediah Strong Smith (1799-1831), one of the earliest American frontiersmen. The biblical name Jedediah originated after the death of King David's first son with Bathsheba. Nathan the prophet (2 Samuel 12:25) added the name Jedediah (Hebrew – Beloved of JAH) to David's next son Solomon, upon his birth to he and Bathsheba. Solomon (or Jedediah) eventually rose to the throne of Israel as king after the death of King David, as promised by God.

Jedediah Strong Smith

Frontiersman Jedediah Strong Smith holds a unique place in the history of the American west. In the early nineteenth century, a few adventurous men headed into the wilderness drawn by the lucrative fur trade. Smith, born in Bainbridge, New York, grew up in Pennsylvania but by 1822 headed to Ohio and then west beyond Missouri to become a successful fur trapper. During one of his early hunting expeditions, Jedediah Smith and a band of frontiersmen on horseback came face to face with an angry grizzly bear.

As I was relating the story, our campfire scene was suddenly interrupted by the roar of a large grizzly bear emerging from the bushes. Promptly, the three frontiersmen guarding our group responded and raised their rifles. They discharged a shot at the bear, killing it instantly. The sound of the black powder rifle discharge startled everyone and a few women in the church screamed at the concussion.

Grizzly Bear Charge

If you had not already guessed, I carefully crafted the scene including a simulated campfire and painted mountain backdrop. One of the church scout leaders wore a rented bear costume for the bear attack scene. The same presentation occurred in the church and during a scout camp outing. Though the shot resounded throughout the building, the black powder discharge fired only a blank paper wad.

After the frontiersmen removed the "dead" bear from the stage, my story of Jedediah Strong Smith continued. When the grizzly bear seized Jedediah Smith, it's claws tore flesh from his chest, exposing and breaking his ribs. The bear commenced to grab Jedediah's head in its mouth, ripping the scalp and hair to the side while tearing off one of his ears.

As the bear stepped back, one of Smith's companions took aim and got a shot before the grizzly could lunge again. The bear fell dead, but Smith was horribly injured with little hope for survival.

Though his companions presumed Jedediah could not recover, he convinced his friend Jim Clyman to take needle and thread and sew up his chest, stitch his scalp in place, and re-attach his ear. After cleaning and treating his wounds, Jedediah soon recuperated and continued his trek west. For the next nine years he led several expeditions across the Sierra Nevada mountains into California, through the Cascade mountains to Oregon, and into the Southwest territory of New Mexico.

During all of his travels Jedediah Smith journaled observations and created maps used by explorers, settlers, and the United States Army for decades. One of his most important discoveries was the South Passage transit route into Oregon territories. In 1831, at age 32, a Commanche Indian raiding party encountered Smith while he was searching for water near the Sante Fe trail and killed him.

The Oregon Trail served as the primary route west for wagon trains of settlers over the next decades. In the following chapter the Oregon Trail features prominently in the story of the Donner Party after their decision to take a "shortcut" to the established route.

Presentation of the Grizzly Bear Charge at a Scout Outing

Notable traits among Jedediah Strong Smith's qualities were a reputation for unwavering determination and a sound commitment to his faith after conversion as a Methodist. Though not having access to places of worship or fellowship with other believers, he maintained a habit of daily devotional reading and journaling. Also, Smith was admired for being temperate concerning alcohol, never swore, and refrained from being alone with women.

Apart from his journals, Smith's biography remained undocumented until seventy-five years after his death. Consequently, there persists some debate about the details of his life. However, he clearly stood apart for his faith from the company of rowdy frontiersmen. Jedediah Strong Smith, unlike most frontiersmen, did not just explore the west to survive and make a living by the fur trade. His journals, discoveries, and maps left a legacy for the future, though he never saw the results in his short lifetime.

Black Powder Rifle, Pouch, and Powder Horn

English statesman, Oliver Cromwell (1599 – 1658), once famously declared to his troops, "Trust in God and keep your powder dry". Before the advent of sealed rifle cartridges, black powder horns were used to protect gunpowder from moisture, lest it be rendered useless in battle. Throughout the era of muzzle loading rifles, Cromwell's saying encouraged faith in God, and constant preparedness for battle. His instruction goes well beyond the soldier in battle.

The story of Jedediah Strong Smith, while historically important, represents one example of universal tribulation experienced by both Jewish and Christian believers. Tribulation denotes not just trials from situations and persons, but more importantly the work of an adversary during conflict. A common feature of life itself includes evil in the form of the occasional opponent; whether person, circumstance, or grizzly bear.

As I explained in the book *Rend the Heavens*, God permits evil as a fitness center for resistance training, personified by the jester as antagonist. God Himself created all other beings, some of which choose by force of will to oppose and to separate from Him. God's adversary provides only resistance, yet he creates nothing. As a result, tribulation for believers transcends every nation and culture throughout history.

Jester in a Bear Costume

Traditionally a jester served royalty in several valuable ways. First, the court clown's function served to humble pompous officials and their wives who took themselves too seriously. He would poke fun at anyone and anything in the palace, though king and royalty were usually off limits.

Second, the jester acted as a "wild card", adding an element of comic relief and surprise to formal events. Finally, when a serious political crisis arose, the court fool created a distraction by arbitrarily assigning blame to deflect criticism from royalty.

Often a jester was small of stature, with outrageous personality, and sporting outlandish attire. Writer Shakespeare labeled him the "Motley Fool". Hence, when a circus act (or life) does not go as planned, the ringmaster or host cries, "Send in the clowns!" Contemporary comedians serve much the same purpose in media, theater, and life. Rather than fatal attraction, the court fool generates an indispensable distraction. In Chapter 12 the history and role of the jester will be addressed in greater detail.

Parable of Ten Virgins

On the heels of His Olivet Discourse (Matthew 24), Jesus spoke several parables, the first of which addressed the kingdom of heaven likened to ten virgins (Greek - maidens). Each went out at night with an oil lamp to meet a bridegroom. Five of them were termed "wise" because they also took a vessel of oil, and five were classed as "foolish" having carried no extra oil.

While the bridegroom tarried longer than expected, they all slept. At midnight a cry went out signaling the groom's impending arrival and a call to go meet him. As the ten prepared to trim their lamps, five foolish noticed their lamps were out of oil. Upon departing to get oil, they missed the bridegroom and the door closed. In their preparation, a primary concern should have been to have sufficient oil in case the wait took longer than expected.

Wise and Foolish Maidens with Lanterns

Many explanations of the parable miss the fundamental concept; prepare for every contingency. "Trust in God ...and keep your powder dry." As we taught our church scouts, prepare for the worst-case scenario. Every event and outing, when confronted by boys lacking necessary provisions, I would ask, "Did you read the list of requirements?"

Often the boy's response was, "No, I didn't read it." Sometimes one would answer, "Yes, but I didn't think it would be necessary." Occasionally, I would hear another say, "Yes, but I didn't do it." Failure on three levels: being but not knowing, knowing but not believing, and believing but not doing. During the days leading to the return of Jesus, these characteristics are exemplified by the church.

In the campfire scene, the story of Jedediah Strong Smith, as well as Jesus's parable of the ten maidens, each story creates compelling theater that explains biblical principles. In chapters ahead we will look at; the script and stage; the cast and critics; the protagonist and antagonist; and finally, the plot and purpose.

As with Christ's crucification and resurrection, despite many twists of the story, the outcome of His second coming is secure. Be prepared!

The mystery of when and how He will come is now unfolding. Watch the signs!

Woman Selling Oil

Old German City of Rothenburg ob der Tauber

Chapter Two - Snowbound

For our 25th wedding anniversary in 1997, Debby and I planned a trip through Europe to coincide with my regular business travels. At the time I worked for the company as a customer representative to worldwide packagers of the LM6000 industrial gas turbine. Since 1989 I had worked on the engineering team to design and develop the LM6000 product.

Late in the year, I scheduled visits with my customer contacts at packagers and operators of the LM6000 engine in several European countries. Most customers use the 40-megawatt (MW) class LM6000 for electrical power generation.

The route of our travels began after landing in Zurich, Switzerland

and coursed through Germany, France, Belgium, and The Netherlands. On the return, we circled back through Germany, Austria, and finally to Zurich again. As usual, I had hoped to do some photography during our travels.

LM6000 Industrial Gas Turbine and Power Plant

Throughout the trip, skies over Europe remained mostly overcast with only a few brief periods of sunshine. Temperatures dropped close to freezing at night but we managed to complete our intended travels and several sightseeing stops despite the chilly weather.

Coming back through Germany we took a tour through the old city of Rothenburg ob der Tauber and stayed overnight in Munich where we visited family before driving up into the Austrian Alps. Departing from Munich, a cold weather front began to move in with potential for rain turning to snow, especially at higher altitudes.

One of my key objectives included a stop at the famous German castle Schloss Neuschwanstein at the foothills of the Alps. In the village of Hohenschwangau below the castle, we stopped to consider our plan in light of the approaching storm. The only way to take photographs at Schloss Neuschwanstein required a climb up a footpath to the base of the castle.

By this time, I had abandoned the objective of a tour due to the timing with potential for snow. Tourists descending from the castle announced the footpath was slippery and the facility would close early.

Neuschwanstein Castle Shrouded in Clouds

At the very least I insisted on walking the footpath to take some pictures. But Debby argued it would risk becoming snowbound in the Alps if we delayed any longer. After a heated debate we compromised and Debby got her way. My consolation included getting a few pictures from the parking lot below with Schloss Neuschwanstein shrouded in clouds.

Timing now became critical. As we climbed the roads into the Austrian Alps in the rain, I noticed the temperature on the dash dropping rapidly to freezing. Suddenly rain transitioned to a mix with large snowflakes and I could make out snow blanketing the mountains ahead. The narrow road showed little evidence of tire tracks through the accumulating snowfall as it increased to a virtual whiteout.

Soon we were in snow approaching the limit of what the car could safely handle. It appeared we either needed to turn around or get onto a road cleared of snow. At a scenic overlook we stopped to evaluate our options. With earlier snowpack and new fallen snow, surrounding scenery displayed several feet of snow with no sign of stopping. While parked, I managed to get a few photographs as the conditions rapidly worsened.

Snow at a Scenic Overlook in the Austrian Alps

Suddenly, to the relief of our predicament, the roaring sound and flashing lights of a passing snowplow sped by the scenic overlook. Quickly we took advantage of the opportunity and departed, positioning ourselves a safe distance behind the plow.

Over the better part of the next half-hour, we slowly climbed to the summit of the alpine road behind the snowplow until the driver exited onto another road. As we descended the other side of the Austrian Alps, the snow diminished until we entered the sixteen-kilometer-long Arlberg tunnel and exited to clear roads down through the Austrian hillside.

Schloss Hotel, Bludenz Overlooking the Austrian Alps

Finally, arriving in Bludenz, Austria, we stayed overnight at the Schloss Hotel overlooking a snow-covered mountain scene. In full moonlight, the Alps glistened beautifully with an eerie glow that did not disclose the danger we had barely escaped. We were blessed with critical timing, good decisions, and a passing snowplow that prevented a snowbound stay overnight on the mountain. The alternative could have been a very different ending.

Snowbound in Donner Pass 1846-1847

After early frontiersmen like Jedediah Strong Smith and Jim Clyman mapped routes west to Oregon and Nevada, wagon trains of settlers began to make their way across the Cascade and Sierra Nevada mountains to the west coast of the Americas. The 1840s experienced a sudden rise in migration west as prairie trails and mountain passages became established routes.

Prior to completion of the trans-continental railroad system, the only alternative to the long trip from the eastern states was a dangerous sea voyage around Cape Horn at the tip of South America. Travel by ship required up to six months or more depending on very unpredictable weather and sea conditions.

Donner-Reed Party Wagon Train to the American West

Likewise, wagon travel usually required as much as six months journey beginning in mid-western towns like St. Louis, Missouri or Springfield, Illinois. Pioneers going from the eastern states would need to travel west to cities called "jumping-off points" at the limit of railroad lines. From there they signed up with a wagon train led by an experienced wagon master.

All travelers were required to have their supplies, oxen, wagon, and personal effects ready to move after the spring thaw in mid-April. Departure later risked being stranded in the high desert or mountains by early snowfall.

Other challenges to the overland route included Indian raids, drought, equipment failure, sickness, and most importantly interpersonal feuds. One group, the Donner-Reed Party (1846-1847), faced all of these obstacles and more, resulting in significant delays. The outcome took a toll on members including a great loss of life. The Donner Party tragedy has become synonymous with the statement of Murphy's Law, "If anything can go wrong, it will."

For the purposes of this book, contributing factors to the Donner wagon train story will only be noted briefly. Problems began in Independence, Missouri with a late start in May. About 500 wagons constituted the original group. Nine wagons of the George Donner and James Reed families were at the rear. Due to additional conflicts and challenges, the Donner-Reed wagons separated and joined a smaller group of 50 wagons. Reed assumed a lead role but his autocratic style conflicted with George Donner and others in the party.

Furniture and Supplies Abandoned Along the Oregon Trail

Once they reached Fort Bridger, Wyoming east of the Great Salt Lake, serious discord began. Lansford Hastings, wagon master of a group on the Oregon Trail ahead of the Donner-Reed party, sent messages back to Fort Bridger recommending his shorter but untested "Hasting's Cutoff", south of the Great Salt Lake.

James Reed chose to take the short southern route rather than the longer northern trip to South Pass on the Oregon Trail. Hastings had traveled the shortcut route on horseback through the Great Basin Desert, but never with wagons.

Jim Clyman, a frontiersman who had stitched Jedediah Strong Smith's wounds after the bear attack, warned James Reed to avoid the Great Basin Desert trail. He cautioned the wagons could not survive the untested passage through the Wasatch mountains. But Reed disregarded Jim Clyman's warning. Part of the shortcut would require clearing their own trail.

That fateful decision resulted in quarrels, loss of wagons and many oxen, as well as Indian raids that caused at least a month additional delay. During a heated disagreement between the ill-tempered leader James Reed and another man in the group, Reed stabbed him to death. By the time the wagon train reached Truckee Lake in the Sierra Nevada mountains it was early November.

Donner Party Memorial at Donner's Pass

A sudden winter storm trapped the Donner-Reed party of 87 persons in what is now infamously named Donner's Pass. Forced to winter in the high mountain basin, only 48 of the party survived. Wild game animals descend to lower altitudes during the winter, so the family members were forced to kill and eat their remaining horses and oxen. By mid-winter, snow accumulation prevented any escape and starvation began to set in. Details of their ordeal until rescue in late-spring defy description and record the worst human conditions.

In an amazing contrast to the difficulties of the Donner-Reed party, travel today through the Wasatch Mountains and Donner's Pass is comparatively easy by overland rail. During World War II, the American Locomotive Company (ALCO) developed the massive 4-8-8-4 wheel "Big Boy" steam locomotive.

Twenty-five Big Boy engines produced were able to pull freight trains across the Wasatch Range east of the Great Salt Lake to California and back without assistance of additional pusher engines. Union Pacific railroad recently restored a Big Boy steam engine number 4014 as a special publicity project.

Union Pacific 4014 "Big Boy" Traversing Donner's Pass

The Donner Party's experience carries a few important lessons. First; the importance of time-based decisions and actions. Remember the ten virgins? Five were there, ready and waiting. When doing contract negotiations for equipment production, we often negotiated around a "Drop-Dead Date" or DDD. That date meant some action must be completed by that timing or there could be no agreement.

The Donner-Reed families allowed too many factors to push them beyond a DDD for crossing the mountains to the Pacific coast. Literally, it ended up being a "drop-dead date". In the parable of Matthew 25, only half of the maidens had prepared for any eventuality.

Second; make decisions based on good advice from a reliable source. God's Word and His Spirit should always be the first guide. Often, I have to accept Debby's intuition rather than my insistence on blazing a new trail. Sometimes the tyranny of the urgent distracts from what needs to be accomplished. As always, an adversary awaits to suggest the wrong path.

Finally; ego, selfish pride, and conflict are the adversary's tools to undermine every purpose of God and man (James 3:14-16). These traits doomed the Donner party. They were also the downfall of Heylel from a position as chief archangel to his fallen role as Satan. But, as we shall see, God promised a closing scene of human history to complete His drama, and His adversary's demise.

Jester Offering a Shortcut

When Jesus delivered His Olivet Discourse (Matthew 24, Mark 13, Luke 21), He referred to a singular event beginning a period of "great tribulation", such as had never before been seen. That event will be a repeat of the tribulations after the "Abomination of Desolation" spoken of in Daniel 9:27.

The coming "great tribulation" stands out against all other historical times and places of human affliction. Troubles suffered in the stories recorded through chapters ahead only offer a foreshadowing of trials yet to come.

> For then shall be great tribulation, such as was not since the beginning of the world to this time, no, nor ever shall be. And except those days should be shortened, there should no flesh

be saved: but for the elect's sake those days shall be shortened."
(Matthew 24:21,22)

Despite all the trouble of this "shortened" time of testing, a period of even greater distress follows in the "...the great and the terrible day of the Lord's wrath" (Joel 2:31). Most of the real affliction recorded in the Book of Revelation occurs after the removal of believers at the sixth seal. The day of the Lord begins at the seventh seal which contains seven trumpet judgments and the seven bowls of wrath unleashed upon the ungodly left behind.

The Mayflower Journey 1620

Chapter Three - Mayflower

THE JOURNEY OF the Pilgrims to New England serves as an example of faith under severe trial. In this case, I have a personal heritage thirteen generations back to one of the key figures, William Brewster (1566-1644). Like Noah's Ark, the Mayflower voyage carries cautionary lessons about the soon return of Jesus Christ.

In order to preface the difficult circumstances leading up to the Mayflower voyage, a short history of their persecution and reasons for the journey are needed. The period of religious reformation in Europe had its roots well before publication of Martin Luther's 95 theses posted in 1517. From 1517 to 1650, growth of the protestant reformation took many forms as the movement distinguished itself from the establishment church throughout European nations and the Americas.

After the vision and conversion of Roman Emperor Constantine in 312 CE, most regional governments and the official church had become intertwined, complacent, and in some cases inseparable. With time, religious expression often turned out to be subservient to the will of government.

In turn, each state governance and their respective church leadership sought to suppress the influence of the reformation movement. Leaders restricted religious worship including sermons and music, while personal reading of the Scriptures was often forbidden, even for those able to read the original languages.

Persecution for non-compliance took the form of fines, property confiscation, flogging, imprisonment, and even execution. The warnings of Jesus Christ to the seven churches in Asia recorded by the Apostle John (Revelation 1:17-3:22) addressed seven significant regional churches of that period. They also represent seven periods of church influence the past two millennia. Finally, these seven characterize church priorities or lack of them in the end times.

Only two named churches in the historical sense, the persecuted church Smyrna and the separatist church Philadelphia, received no rebuke. Smyrna, representing the Formation church of the second to fourth centuries, suffered greatly at the hands of the Roman Empire as well as Jewish leaders. The church of Philadelphia, spanning the sixteenth through eighteenth centuries, represents the Re-Formation church period including the Pilgrim movement. Both experienced intense tribulation and martyrdom, yet maintained a true profession of faith.

Portrait of Theodore Beza (1519-1605)

One of the key themes of this book addresses universal suffering of believers. When tribulation raises its head, the message of Theodore Beza (1519-1605) bears remembering. Beza, a French Calvinist theologian, wrote a defense of the reform church to King Henry of Navarre (modern day France) before the massacre of French Huguenot believers. The purge resulted in the death of an estimated 30,000 Huguenots on St. Bartholomew's Day 1572. In his letter to King Henry, Theodore Beza concluded his case to the King with an appropriate summation.

> Sire, it belongs to the Church of God, in the Name of Whom I speak, to receive blows, and not to give them, but it will please

> your Majesty to remember that the Church is an anvil which has worn out many a hammer. (Theodore Beza, 1562)

Throughout history the following claim of Jesus Christ has been verified each time believers faced life threatening persecution.

> And I say also unto thee, That thou art Peter, and upon this rock I will build my church; and the gates of hell shall not prevail against it. (Matthew 16:18)

Jester with Hammer and Anvil

The gates of hell signify a defensive strategy of the adversary, not an offensive position. Every time Christian believers seek to renew the church back to its biblical roots and reform the culture to righteousness, the adversary resists by persecuting believers. To follow the metaphor of Beza, suffering does not destroy the anvil, but blows eventually wear out the hammer.

The Puritans

Puritans represent one important expression of the reform movement in England and the resulting oppression. When William Tyndale (1494-1536) translated the Bible from original languages, he predicted a boy behind a plow would know more of the Scriptures than the pope. For this Tyndale was betrayed and burned at the stake. On the stake he cried out, "Lord, open the King of England's eyes!"

Portrait of William Tyndale (1494-1536)

Eventually King Henry-VIII did authorize the printing and reading of the Bible but with the intent to undermine papal authority. Because the pope had refused to approve King Henry's divorce from Catherine of Aragon and remarriage to Anne Boleyn, the King split from the Roman Catholic Church and established the Church of England in its place (1534).

After the death of Henry-VIII in 1547, his son Edward-VI assumed the throne of England. During a period of tolerance toward reform in the Church of England, the class of "purifiers" eventually assumed the name Puritans. Freedom for reform ended upon the death of King Edward-VI in 1553 and succession of Queen Mary who received the nickname "Bloody Mary" due to her purge of church reformers.

Mary did not approve of the Church of England split and desired to reinstitute Catholic practices. Louis Carroll, in his story *Alice's Adventures in Wonderland* (1865), claimed he did not specifically parody Queen Mary by the Queen of Hearts crying, "Off with their heads!" But the caricature fit.

During five years of Queen Mary's reign, many Puritans fled to Europe, particularly Holland and Geneva. There they were relatively free to study and practice their faith. During this period, they adopted the Geneva Bible, the first translation to include divisions of chapter and verse.

Upon the death of Mary, Elizabeth-I rose to the throne in 1558, commencing the Elizabethan Era. With a more favorable government, Puritans who fled from England returned again to their homeland. With them came the popular Geneva Bible and a greater understanding of biblical teaching.

By 1562, Queen Elizabeth-I initiated prohibitions on reform to limit calls for further change. This led the Puritans to separate themselves from the Church of England, whereupon they were labeled "English Separatists" and in 1580 they created a distinct movement. However, increasing religious, political, and social pressure moved the Puritans to seek for opportunities to establish their own community. Their search led them to embark on a voyage to the New World of the Americas where they would face a whole range of new challenges.

The Mayflower Under Sail 1620

Voyage of the Mayflower

In the town of Scrooby, North England, Separatists founded a church based on a covenant of self-government. The year was 1606 and several important leaders arose to guide the community. Three well-known names are William Bradford, William Brewster, and John Robinson. William Bradford, future governor of Plymouth colony for 33 years, became recognized for his 1647 manuscript of *History of Plymouth Plantation*, documenting the founding of the colony.

Equally well known was William Brewster, who, as elder and pastor had a background of influence within the court of Queen Elizabeth-I as personal secretary to Sir William Davison. According to William Bradford, Brewster established himself as an effective evangelist. Less well known, John Robinson served as a pastor and Christian philosopher during the time in England and Holland. When the Pilgrims departed on the Mayflower, members of the Separatists who remained in Scrooby urged Robinson to stay on as pastor of the congregation and delay travel until a later voyage.

The decision to make the passage to colonial American and preparations for transportation included many trials. Most of the congregants lacked sufficient means to help underwrite the cost of the trip let alone manage the logistics of approval from government bureaucracy in Holland, their original point of departure.

Obtaining a suitable ship, negotiating the merchant company endorsement, and securing an official "patent" to approve the trip presented significant obstacles in their own right. Published stories of the Mayflower journey include extensive details of their challenging preparations and final departure in July to England and finally on to New England.

Jester Short-selling the Speedwell

When the Mayflower and sister ship Speedwell set sail from Southampton, England in August 1620, they were forced to return due to leakages on the Speedwell. Even after repairs, a second attempt resulted in leaks so severe the Speedwell had to be declared un-seaworthy.

After transferring some of the passengers and cargo from Speedwell, the overcrowded Mayflower made a final departure for the Virginia Colony on September 6, 1620. Of the 102 passengers, most were classed as "strangers" to the Pilgrim community, traveling to the New World as crew, for adventure, or financial motives.

On November 9 the harrowing journey ended with sighting of land at Cape Cod, as a consequence of being blown off course in a storm. Upon arrival they first joined to give thanks to God for safe passage. Their attempt to sail on to the original destination of the Virgina Colony met with continued unseasonable weather, and they finally returned November 21, resigning that Divine providence had placed them in Cape Cod. By then winter set in with all New England's fury.

Signing of the Mayflower Compact (1620)

Mayflower Compact

While still on board the Mayflower, forty-one persons including Pilgrim leaders, congregants, and some of the "strangers" gathered to join in a compact. The framework followed the agreement in Scrooby forming a covenant of church self-government. The Mayflower Compact, based on biblical principles, later served as a foundational document for democracy and eventually the Constitution of the United States.

This approach of binding together for mutual benefit and self-governance departed from traditional rule guided by the concept of the "divine right of kings". By a covenant, the governed agreed to be bound in good faith, professing that true authority to rule comes from God Himself, rather than through a ruler. A group of elders, identified by their biblical understanding and experience, served as leadership for the church and the community.

The first winter proved unforgiving to the settlers. A lack of adequate shelter, limited food resources, and disease took a heavy toll, resulting in the death of almost half of the pilgrims. By March of 1621, only 55 of the original 102 remained. Several years prior to their arrival the local village of the Patuxet tribe had been virtually wiped out by an outbreak of disease. In the clearing that remained, the small group of colonists erected a makeshift settlement.

Squanto and the Native Tribes

Then, in a sequence of improbable events, two encounters with the native tribes rescued them from certain devastation. The first named Samoset appeared with a greeting, "Welcome Englishmen!" A survivor of the Patuxet tribe, he had worked with English fishermen as far north as Maine, and learned to speak their language fluently. Shortly after, Samoset appeared with another visitor Tisquantum, whom they named Squanto.

Squanto not only spoke English, but he had been kidnapped at age twelve and taken to England for nine years, where he became well versed in their customs, beliefs, and language. Upon returning to his native home, Squanto suffered a further kidnapping and journey to Spain until he was rescued and ransomed by Spanish monks. Again, he obtained passage back to New England.

Squanto Teaching Pilgrims to Sow Corn

Over the next year Squanto provided invaluable lessons to the Pilgrims, showing them how to efficiently harvest fish and eels, as well as how to properly plant corn with a fish for fertilization. In April Captain Christopher Jones and the surviving crew members chose to return to England in the Mayflower. Despite an invitation to the Pilgrims to join the voyage, they all chose to remain and continue in Plymouth.

Soon the Pilgrims had twenty acres sown with corn, providing the harvest for the first Thanksgiving celebration that fall. At the invitation of the settlers, many local tribesmen joined and enjoyed themselves so much they continued the feasting for three days.

Thanksgiving Gathering of Pilgrims and Native Tribesmen

Sadly, in 1622 Squanto contracted a fever and passed away, at what time he made a profession of faith in Jesus Christ. Squanto's interaction with the Pilgrims of Plymouth Colony turned out to be mutually beneficial. William Bradford declared Squanto was, "a special instrument sent of God for their good".

In 1630 a large group of about a thousand Puritans sailed to New England forming the Massachusetts Bay Colony. These Puritans, were not "Separatists" and remained within the Church of England. Seventy-one years after the arrival of the Pilgrims, Plymouth Colony merged into the Massachusetts Bay Colony. The years and decades that followed were not without struggles. Plymouth's community of Pilgrims dealt with improper governance, financial mismanagement, and an occasional conflict with the Indian tribes.

Collective Starvation

The Pilgrims fled England and Europe not only for religious freedom but also to escape the seventeenth century feudal system. In the European model of land and labor, property belonged to the monarchy, aristocrats, and the church. A peasant class worked tracts of land owned by aristocrats for a portion of the harvest while skilled laborers and artisans performed work enforced by rules of communal trade guilds.

Formation and propagation of trade guilds were initially resisted by the medieval church due to the use of secret oaths, obligation to support members above all other allegiance, and the depravity of drunken celebrations. Gradually the clergy integrated fraternal trade associations into religious practices through the adoption of patron saints and payment of a mandatory guild tax to support church construction and operation.

Medieval guilds regulated commerce through the pricing of goods and services, setting standards for quality, and restriction of competition from non-members. Communal services such as welfare and funerals funded by guild dues provided a safety net of mutual aid for members.

Within this framework of guild culture, the Pilgrim's contract agreement with English financiers of the Mayflower project required operation as a communal society. In order to assure that sponsors back in the Old World received a share of benefits, they insisted on shared property and collective harvests. Supporters based their fear on the assumption that Pilgrims would accumulate personal wealth not subject to their backer's assessment tax.

Therefore, each person received equal distribution of the harvest whether or not they contributed labor. One of the transformational changes to Pilgrim culture in the Plymouth colony involved suspension of communal responsibilities. By 1623, several years of shared farming demonstrated the failure of a communal model. In the collective system, parity resulted as everyone suffered equally from a lack of incentive. As a result of quarreling, starvation, and widespread depression, in the spring of 1623 governor William Bradford ended collectivist farming.

With counsel of fellow leaders, Bradford divided the farmland property among families. The result of tying effort to output created an immediate improvement in production as well as general cooperation and well-being of the members. Abundance meant everyone benefited, including their financial sponsors.

Risk and Reward

In the book RtH, I related how Debby and I worked in the ministry of an organization World Ministry Outreach (WMO) for four years (1971-1975). During that time working in Haiti and Florida we received no pay from the organization. WMO met our expenses through the contributions of participants who held paying jobs.

Like many startup ministries during the Jesus Revolution, WMO operated on a simplified collectivist principle; equal benefit regardless of effort. Like all collective experiments; common belongings in the early church, a Pilgrim communal model, and WMO's shared benefit, each failed due to inherent abuses. As with the parable of talents (Matthew 25:14-30), they collapse because the motivation of risk and reward is removed.

William Brewster

William Brewster's Legacy

From an early age, I recall my mother Letah Stroh McGreehan (1921-1975) relating details and stories of her family ancestors. Named Retah Pearle Stroh at birth, she was the last of four sisters, born to a German-English family in Batavia, New York. Upon entering school, her parents changed the name Retah to Letah in order to curb misspelling of her first name as Rita. Phonetic writing of names, common in the nineteenth century, often carried over into the early twentieth century.

With the exception of the lineage of very prominent persons, most readers are uninterested in hearing details about family history. Even for well-known celebrities, a list of names and dates with "who begat whom" strains the attention span. However, readers can relate when names, dates, and begets are woven into a story that provides meaning to historic or current events.

Whether it concerns mathematics, science, literature, or history; I often hear the expression, "That is a boring subject." G.K. Chesterton (1874-1936), an influential Christian apologist and writer once penned, "There are no uninteresting things, only uninterested people." Chesterton's observation will become relevant as the importance of names, dates, and stories are tied to the theme of this book.

Gravestone, Amos Spring, Sr (1752-1814)

Our personal investigation into family history began with Debby's project to chronicle personal accounts from her many aunts and uncles. After years of visits to distant relatives, cemeteries, and libraries, she managed to fill the equivalent of two large filing cabinets with her genealogical history. In the process, her research uncovered records of family members back as far as the early 1700s. That included names of several soldiers that served in the Revolutionary War (1775-1783). Occasionally, Debby would repeat a popular refrain, "An old person is a library on fire."

Amos Spring, Jr (1779-1850)

The New England Spring-Snow Family

After my mother passed away in 1975, I inherited a copy of her family history prepared by her mother Ruth Spring Stroh, and her grandmother Albina Chaddock Spring. Not only did it include a comprehensive lineage of their immediate families, but extended back a total of twenty-seven generations to the twelfth century. Among them were seven eighteenth-century enlisted men and officers in the Continental Army during the Revolutionary War (1775-1783). One of my seven Revolutionary War ancestors was Amos Spring, Sr (1752-1814), identified on my application to the Sons of the American Revolution. The next chapter covers important lessons from that conflict.

Darius Newton Spring (1817-1896)

It should be of note, recorded genealogical heritage is only as good the research behind it. Even though my grandmother Ruth Spring Stroh and her mother Albina Chaddock Spring meticulously investigated and documented her family history, some of the information prior to the sixteenth century was supplied by their paid researchers.

Fires, floods, and wars, often resulted in Early American and European records lost or destroyed. In at least one case we determined a family birth record had been falsified to cover embarrassing events. A lack of literacy often meant information was not recorded, or difficult to interpret.

David Scott and Albina Chaddock Spring

Particularly interesting was my family ancestry back to Elder William and Mary Brewster on the Mayflower. Not merely were names and dates provided, but documentation included personal stories of early settlers as well as their role in contemporary events.

Pilgrim Tribulation

Unlike the mostly peaceful cooperation of Plymouth Pilgrims with local tribes, many New England governments including Massachusetts Bay Colony retaliated against Indian raids by issuing bounty payments for Indian scalps.

Gradually settlers in New England moved west to wilderness in what is now Upstate New York. Land Offices received large grants in unsettled regions to sell with the objective of enticing families to clear and develop the land.

Prior to the War of 1812, Amos Spring Jr with his wife Reliance Snow Spring, six children, and his brother Darius moved from Massachusetts to the unsettled area around LeRoy in Western New York. Traveling by ox drawn wagon with fifty cents in his pocket, Amos and Reliance Spring cleared land and built a home in what was then wilderness.

Albert and Ruth Stroh and Daughters, Evelyn, Gladys, Verna Mae, and Letah

The Biblical story includes a Hall of Fame in Hebrews 11 listing men and women who, by faith, endured great affliction, torture, and even death for their witness. Paul the Apostle (likely author of Hebrews) recorded,

> These all died in faith, not having received the promises, but having seen them afar off, and were persuaded of them, and embraced them, and confessed that they were strangers and pilgrims on the earth. (Hebrews 11:13)

As with each of the stories in this book, through Old and New Testament, as well as the culture, most great accomplishments result from great affliction.

Minutemen at Lexington and Concord

Chapter Four - Revolution

ALMOST ONE HUNDRED fifty years after the arrival of the Mayflower at Plymouth, Massachusetts, the colonies experienced an agonizing rift with the English crown. By the end of Britian's Seven Year War (1756-1763), France relinquished to England all claims in the American Colonies. After the signing of the Treaty of Paris in 1763, most residents in the colonies proudly embraced English citizenship. This sentiment changed dramatically over the next ten years.

In order to further consolidate control by the British Crown and support the need for resources, the decade following 1763 saw King George-III (1738-1820) refuse to grant further autonomy to the colonies. The Stamp Act of 1765 increased tensions by a mandated taxation stamp affixed to all documents and items sold. Thus began a forty-year period of conflict starting with the Revolutionary War (1775-1783) and ending after the War of 1812 (1812-1815).

To encourage the colonists, Thomas Paine opened his patriotic booklet *The American Crisis* (1776) with the line, "These are the times that try men's souls." Particularly during the next eight-year war, most of the new nation's founding fathers, as well as soldiers, financiers, and the general population suffered greatly.

Rather than religious freedom, colonialists sought self-determination and political liberty. Indeed, the pursuit of those objectives over those forty years would bring the government, economy, and social fabric to the brink of collapse. In the midst of great personal sacrifice, the spiritual climate of the emerging nation grew stronger. How ironic that the greatest tests of the nation have fostered spiritual renewal.

The following selected events illustrate God's providential help experienced against seemingly impossible odds. In fact, so many improbable occurrences took place that among colonialists, the term Providence came into common use interchangeably as a name or title for God.

As I explained in the book *Rend the Heavens ...and Come Down*, the word providence derives from the Latin *pro-videre*, or to look ahead and prepare. As Abraham declared on Mount Moriah after finding a ram caught in the thicket, "*YHWH jireh*" (Yahweh provides). God planned ahead on that same Mount Moriah with Jesus as the sacrificial Lamb.

Colonialists Throwing Tea Boxes from British Ship

The War Started by a Party

Tensions between colonists and England escalated with the refusal to pay an imposed tax on imported tea. Rather than send back a tea shipment that arrived at Boston Harbor in December 1773, a group of patriots disguised themselves as native Americans and proceeded to dump the tea overboard into the harbor. That action referred to as the "Boston Tea Party" constituted a shot across the bow to the British Crown regarding "taxation without representation".

Fourteen months later on April 19, 1775, after a series of legislative reprisals by the British Parliament, the first armed conflict took place at a bridge near Concord, Massachusetts. Seven hundred British Regulars marched toward Concord under secret orders to confiscate colonial military supplies. After an early skirmish, fighting broke out between 400 militia and 100 British Regulars at the Concord bridge.

As the Redcoats retreated back to Boston, militiamen exchanged fire along the route. By the next morning, Boston came under siege by a gathering of 15,000 militiamen. When word arrived to George Washington at his Virginia home of Mount Vernon, he declared, "...can a virtuous man hesitate in his choice?" Washington received the call from the Continental Congress as Commander-in-Chief of the army on June 15, 1775. He assumed command in Philadelphia, Pennsylvania.

Washington Assuming Command of the Army

George Washington Called to Lead the Army

By this time, the reputation of George Washington preceded him after serving with the British army during the French and Indian Wars. During Washington's time as aide to General Edward Braddock, he fought in the campaign of 1755 against French Fort Duquesne in what is now Pittsburgh.

A surprise ambush by French and Indian defenders managed to rout a much larger English force of 1500 regulars. Of about a hundred officers, all were killed or wounded except George Washington, who served as an officer with the colonial militia. Captured British soldiers were scalped and burned alive by the Indians.

During the British retreat from Fort Duquesne, the Indians fought from the woods along the roadway, killing two-thirds of the Redcoats. The Indians under Seneca Chief Guyasuta targeted George Washington as his six-foot two-inch height in a uniform made him an easy mark. Washington had two horses shot out from under him, as well as his hat and coat pierced by several bullets, none of which so much as grazed his skin.

Washington Targeted During French and Indian War

George Washington's companion and physician Dr. James Craik, recorded an encounter they had with Seneca Indian Chief Guyasuta. In 1770, fifteen years after the Fort Duquesne battle, they met Guyasuta on a trail near current day Pittsburgh. The old chief approached Washington and declined to shake his hand.

Instead, out of respect the chief bowed before Washington and recalled how he had admired the colonial officer's courage in battle. Despite his Indian braves being so close they could not miss, the chief said, "...Twas all in vain, a power mightier far than we shielded him from harm".

In the recollection of Dr. Craik, the chief proceeded to speak prophetically that Washington would not die in battle but become a chief of nations. George Washington himself marveled at the supernatural way he often escaped death or capture. Writing after the Battle of Monmouth, New Jersey, Washington recalled, "...the hand of Providence has been so conspicuous in all this that he must be worse than an infidel that lacks faith."

Colonel Knox Transporting Cannon from Fort Ticonderoga

In January 1776, Washington ordered Colonel Henry Knox and his men to transport sixty tons of artillery over frozen ground from Fort Ticonderoga, New York. The fort had been seized from the British in May 1775 by Ethan Allen and his band of militiamen. Knox managed to move the cannons using oxen teams and sleds over 300 miles of rough terrain to Boston.

Arriving in Boston March 2, 1776, General Washington created bulwarks and twenty cannon placements overnight on Dorchester Heights overlooking the harbor. Unseasonably favorable weather permitted the artillery to be moved at night and fortified before dawn. The amazing accomplishment took place only after a heavy fog blanketed the bay and British warships below, concealing the fortifications of Dorchester Heights.

Washington's Fortification of Dorchester Heights

Upon seeing the gun placements and occupation on Dorchester hill in the morning, British General William Howe decided to attack the following night. He positioned warships in the harbor for off-loading of additional assault troops. Late that night the weather suddenly reversed as the winds rose to hurricane strength. Driving sleet caused the hill of Dorchester to become too slippery to negotiate. Two of the British troop ships blew ashore and grounded.

General Howe reported the sudden change in weather as the reason for calling off his attack. He offered to George Washington that the British would depart Boston without burning the city provided they were given safe passage to evacuate. The following Sunday Reverend Abiel Leonard spoke of the providential event by referencing the Book of Exodus, saying the Lord,

> "...took off their chariot wheels, that they drave them heavily: so that the Egyptians said, Let us flee from the face of Israel; for the Lord fighteth for them against the Egyptians. (Exodus 14:25)

Washington's Escape from Long Island, New York

By late August 1776 George Washington and a large contingent of the Continental Army occupied Long Island, originally called the village of New Amsterdam. General Howe recognized the opportunity to swiftly surround and capture the city while simultaneously dispatching Washington's forces. To implement the plan, Howe assembled thirty-two thousand English soldiers and Hessian mercenaries, the largest army the British ever amassed on foreign soil.

Though Washington felt reluctant to abandon Long Island, a fearful storm the night of August 21 cemented the decision to leave before the expected British invasion. The next morning the British began their assault from Staten Island toward the American forces on Brooklyn Heights. Howe met initial success by killing or capturing a portion of Washington's men, including three American generals.

Washington assembled his remaining force of nine thousand men near the East River facing a mile of open water to the west. That evening a sudden storm soaked the army, but at the same time prevented British warships from moving to prevent their escape. At Washington's orders every available boat was requisitioned to ferry the army across the East River to Manhattan.

By the break of day, ten crossings had been completed and the final regiments of men waited for the transit boats to return. As if to conceal their getaway, a dense fog developed and remained until the last of the men escaped. Upon disembarking on the Manhattan side of the river, they observed the British suddenly overtaking the empty fortifications, while vainly discharging their muskets. Everyone involved marveled at the miraculous timing that enabled a successful escape.

Battle of Trenton, New Jersey

By Christmas 1776, an army of Hessian mercenaries occupied the city of Trenton, New Jersey. Faced with many of the men nearing the end of their enlistment, George Washington badly needed an offensive victory to boost their morale. He chose to cross the Delaware River in the worst of winter conditions for a surprise attack on Trenton.

By this time most of the Hessian men and officers were enjoying Christmas festivities. A warning note from a British sympathizer to the Hessian commander Colonel Johann Rall went unnoticed because he could not read English. Rall slipped the note in his coat pocket during a card game.

As the crossing began, a sudden blinding snowstorm moved in. Though it created miserable conditions for the American troops, the storm caused Hessian sentries to seek cover as well as concealing Washington's boats and men. In the sudden attack a thousand Hessians were taken captive.

Divine providence again protected Washington's men. Colonel Rall, mortally wounded during the attack, regretted not having given attention to the warning the night before. Had he done so and prepared, Rall claimed before passing away that the outcome would have been quite different.

Washington's Army in Valley Forge

Winter in Valley Forge

The winter of 1777-1778 put the Continental Army in the worst imaginable conditions, yet with a most favorable outcome. Forced to retreat to Valley Forge, twenty miles outside of Philadelphia, General Washington and his troops endured severe winter weather lacking food, shelter, and adequate clothing. During this test of their stamina, help came in several providential ways.

Baron Friedrich Wilhelm von Steuben, a Prussian army officer, began his military career in 1744 by enlisting in the Prussian Army. With officer training von Steuben rose to the rank of captain but was discharged in 1763. In a civilian capacity he received an honorary title of baron, but sought another military role.

Baron Friedrich Wilhelm von Steuben

By 1775 von Steuben received a recommendation for a Congressional appointment from Benjamin Franklin, acting minister to France. With the approval of General Washington, von Steuben was selected as an inspector to the Continental Army. Upon arrival at Valley Forge, Pennsylvania in February 1778, the lack of hygiene, basic discipline, and military preparation appalled Baron von Steuben. Immediately he established a strict plan for kitchens and sanitation.

To improve on administration, von Steuben insisted on orderly bookkeeping and records. Most importantly, he began by instructing a group of 120 trainees on military drill. This process involved appointing successful trainees as instructors of their own regiment, and so it progressed. Von Steuben's commanding demeanor and colorful language gained the immediate respect of Washington's troops.

Baron von Steuben eventually completed a training manual called the "Blue Book" for American troops which continued as a guide to United States Army training for many decades. His influence on the effectiveness of George Washington's Continental Army demonstrated immediate results with victories at the Battle of Barren Hill in May 1778 and the Battle of Monmouth in June 1778.

Washington's Logistical Challenge at Valley Forge

While Napoleon Bonaparte has been credited with saying, "An army marches on its stomach.", the reality is much broader. No conflict can endure without financial backing to provide personnel, arms, fuel, food, clothing, and equipment. Effective leaders recognize the importance of procuring resources and getting them supplied to the frontlines; commonly called logistics.

A fundamental principle of warfare is, "Never outrun your supply lines." That applies to personal, professional, and most importantly, spiritual resources.

Financiers of the American Revolution

Continental Cash Crisis

The Continental Congress faced a challenge of funding for Washington's army. Congress had no central government structure and no established means to secure financial capital. Maintaining enlistment of army personnel required assurance they would be paid for their service.

To prevent desertion from the ranks, George Washington could not depend on authorizations from Congress, and often relied on sympathetic civilians. Congress and General Washington benefited from several very successful financiers. The three most prominent of these men were Haym Salomon, Isaac Moses, and Robert Morris.

Salomon and Moses made their fortunes in banking and as financiers while Morris became a successful businessman and merchant. Each understood the concept of economics as opposed to money. It bears remembering, "Money never solves a financial problem."

Two Jewish financiers with access to international commerce, Haym Salomon and Isaac Moses, enabled them to turn bills of exchange into currency for Washington's army. When a shipment of raw materials left the colonies for Europe or finished goods departed Europe, a bill of lading would be created as a guarantee against a very large sum for the cargo. While in transit, the bill served as a security to cover the risk of loss, and as such, it became a negotiable instrument. These could be bought and sold or used to secure a line of credit.

Isaac Moses and Robert Morris served the fledgling administration by setting up a system of banking and finance for the colonies, often securing loans from Dutch Jewish lenders. One of Robert Morris's contributions included stopping the printing of Continental currency to cover debt, thus preventing hyper-inflation.

Each of these men accepted great personal financial risk as well as potentially death for rescuing the colonies from certain economic ruin. As Benjamin Franklin proclaimed, "We must all hang together, or most assuredly, we shall all hang separately."

Surrender of General Cornwallis at Yorktown

French Alliance and British Surrender

Without the entry of France in support the American Colonies, the Revolutionary War would have been prolonged or, at worst, experienced a different outcome. Based on the influence of Benjamin Franklin, minister to France, the French Foreign Minister Comte de Vergennes and King Louis-XVI were convinced to join the American Revolution. After the successful Battle of Saratoga, Franklin was able to negotiate the Treaty of Amity and Commerce as well as a Treaty of Alliance in February 6, 1778. Lacking a meaningful navy, the colonies needed France's naval support to counter British dominance at sea.

By 1776, France had already begun to provide rifles, gunpowder, and supplies to the Revolutionary Army. But the active entry into the conflict in 1778 resulted in 6,000 French soldiers added to support the Americans. At the Battle of Yorktown, the timely arrival of thirty-six ships of the French Navy in the Chesapeake Bay on August 30, 1781 prevented the British Navy from evacuating the army of General Charles Cornwallis.

With no avenue of escape, General Cornwallis officially surrendered his army October 19, 1781 to General Washington at the Battle of Yorktown. Cornwallis having fallen ill on that day, sent his second in command, Brigadier General Charles O'Hara to offer the sword of surrender. In turn, General George Washington had Major General Benjamin Lincoln take his place to accept surrender. Final cessation of hostilities occurred with the Treaty of Paris on September 3, 1783.

The Constitutional Convention

Formation of the Republic

The title of Shakespeare's play, *All Is Well that Ends Well* has become a popular euphemism. As much cannot be said for the founding of the United States after the Revolutionary War. Despite a successful conclusion and treaty with England, the real challenge lay ahead.

A Declaration of Independence, though effectively calling for a divorce from England did not define the new nation in makeup or operation. Articles of Confederation, drawn up in 1776 through a committee led by John Dickinson, suffered from glaring deficiencies.

Divorce does not solve a problem; it only creates a different set of problems. With the Treaty of Paris ending the war, the United States independence faced a new set of problems demanding resolution.

Rising discontent with failure to pay veterans of the war as well as foreclosures due to unpaid debts generated the seeds of revolt. In fact, several near riots such as the Shay Rebellion of veterans almost spread into wider chaos.

In order to establish founding principles and define a functional government, a Federal Convention convened May 25, 1787 and finished September 17, 1787. With George Washington officiating, the convention delegates met for the stated purpose of revising the flawed Articles of Confederation, but by consensus they determined the need to take a fresh approach.

As negotiations progressed, state delegates' position on issues such as proportional population representation verses equal state representation generated such conflict it appeared there could be no agreement. Another issue dividing the delegates involved the slave trade, which they addressed by putting off the matter of slave importation until 1808. On several occasions participants feared the colonies would separate into factions if they did not reach a consensus.

Just when the Convention came to the point of dissolving over disagreements, Benjamin Franklin stepped in to call the body back to faith in God's providence for a successful outcome. Even to that advice the delegates could not agree.

A Constitution and Bill of Rights

Finally, through a series of contentious votes and stalemate, the delegates found accommodation that broke the deadlock. To balance and limit the powers of government they established three equal branches: Executive, Legislative, and Judicial. In the matter of states' representation, they created a Senate with equal representation and lower House of Representatives with proportional membership.

As the weather cleared and the logjam of disagreements dissipated, a final Constitution and Ten Amendments or Bill of Rights emerged. Even the critics marveled at the apparently miraculous way in which the documents were drafted.

Despite some opposition, all of the delegates signed the Constitution and over time, all thirteen original colonies finished ratification. George Washington, who presided over the negotiations, offered his view that the final result was a miracle.

Jester and the French Reign of Terror

The French Revolution and Reign of Terror

Reflecting on the providential hand in the American Revolution, the pattern of events can be easily misinterpreted. Success in the birth of a functioning republic depended on a populace seeking divine guidance. The nation at its core was deeply spiritual and biblically literate, notwithstanding it's many flaws. In his speech to the Virginia

Convention on March 23, 1775, Patrick Henry closed with the declaration, "Give me liberty or give me death". Patrick Henry's sentiment resolved that liberty made life livable.

An attempt by the French to replicate the miracle of the United States, instead played out on an international stage that degenerated into chaos called the Reign of Terror (September 1793 to July 1794). After the successful American alliance with France and growing dissent of the common people with the French aristocracy, a revolution appeared to be inevitable. However, the French Revolution set off on the wrong foot trying to right the nation.

Failure of the French economy became the tipping point. Lacking a biblically grounded populace, the rebellion focused on a number of erroneous humanistic principles. The August 1789 Declaration of the Rights of Man proclaimed humanistic ideals of personal liberty, parity before the law, and the brotherhood of man.

French Motto of the Reign of Terror

The motto of the French Revolution declared, "*Liberte', Egalite', Fraternite', ou la mort*". The sentiment "*...ou la mort (or death)*" was dropped after the Reign of Terror, because it became interwoven with the execution of aristocracy and the opposition. That emotion departed significantly from Patrick Henry's intent during his speech.

Indeed, for anyone not in step with the revolution came a wagon ride to the guillotine. Failure of the financial system led to the collapse of agriculture under the feudal system of propertyless serfs.

Intellectual Sophists and the Private Salons of Paris

A Cult of Reason Abolishes the Church

Driven by the concept of the Age of Reason, the common people of France began to call for social and political change in France. Pressure came mainly from the popular Salons of Paris rather than from a godly pulpit. Through the late 1600s and the 1700s, many of the aristocratic Parisian women hosted guests by invitation in private salons. These informal gatherings sought to foster diverse interests in art, philosophy, sciences, and politics. By discussion and debate, these influential women increasingly guided public opinion.

From the Salons the seeds of discontent spread, fueled by radical social views of philosophers such as Voltaire [Francois-Marie Arouet] (1694-1778). Voltaire's philosophy against authority and religious dogma led to a cry to silence the already ineffective church.

Gradually the Salons devolved into an anarchistic movement against religion and the church. Closure of the Dominican Monastery of Saint Jacob in 1790 gave rise to the Jacobins, a shadow holy league against the enemies of liberty. Inspired by lawyer Maximilien Robespierre, the Jacobins incited hostility undermining the aristocracy and church.

As the Jacobins campaigned for a Cult of Reason and abolition of Christianity, churches were transformed into Temples of Reason and church property confiscated. The misnamed Committee of Public Safety introduced the title of *citoyen* (citizen) rather than *monsieur* (sir) and a new form of dress adopted as peoples' attire. Also, a new monthly chronology of three ten-day weeks replaced the customary calendar.

The call by Robespierre for elimination of the aristocracy led to adoption of the guillotine for swift execution. Over the following months, up to seventeen thousand people from all classes were beheaded during the Reign of Terror, ending with the execution of Maximilien Robespierre himself. Napoleon Bonaparte, arrested as a revolutionary general with the Jacobins, later received a pardon upon restoration of order.

Napoleon Bonaparte Crowning Himself Emperor

The elevation of Napoleon to power after the Reign of Terror provides a foreshadowing of how the Antichrist will rise from obscurity to seize power throughout the world. At his coronation, Napoleon Bonaparte notoriously took the crown from Pope Pius-VII and crowned himself Emperor of France to demonstrate that his power came from his own achievements rather than God or church. He managed to lead the French army conquering most of Europe before humiliating defeats in Russia, the Battle of Waterloo, and final exile to the island of St. Helena in 1815.

Jester as Napoleon Bonaparte

Consequences of Haitian Independence

Simultaneous with the French Revolution, the uprising for independence in the Caribbean nation of Haiti greatly impacted expansion of the United States. In 1791 a slave rebellion broke out in the French colony of Saint-Domingue, on the present-day island of Hispaniola. At the time Saint-Domingue generated more revenue for France through the sugar cane trade than all of its' other colonial territories combined.

A revolt began August 21, 1791 during a secret gathering of thousands of slaves who assembled for an African voodoo ceremony. Participants viewed the sudden rise of a violent tropical storm as an omen favoring an uprising. As the rebellion expanded, rebels led by ex-slave Toussaint Louverture attacked and killed many French aristocracy and landowners along with their families. Haitian rebels paraded through the streets displaying the severed heads of aristocrats' children during the rioting that followed.

France and England sent military expeditions to the island to suppress the uprising but both resulted only in the death of hundreds of thousands of slaves and militia members. After Napoleon Bonaparte granted independence to the nation of Haiti in 1804, the new president General Jean-Jacques Dessalines ordered the execution of all remaining whites on the island. In my book *Rend the Heavens*, I document personal experiences working several years in Haiti and the struggles of the people to sustain a successful government and society.

Faced with financial pressures and the inability to fulfill the French dream of a Louisiana colony in the Americas, Napoleon Bonaparte agreed to sell the Louisiana Purchase. In 1803, President Thomas Jefferson arranged to buy the Louisiana Purchase for fifteen million dollars, thus doubling the area of the United States.

Like Napoleon Bonaparte, many significant figures rose to power suddenly, offering a preview to the coming rise and fall of the Antichrist. The characteristics of the Antichrist will be covered in detail in Chapters 12 and 17 of this book. Paul the Apostle, speaking of the coming return of Jesus Christ, warned believers,

> Let no man deceive you by any means: for that day shall not come, except there come a falling away first, and that man of sin be revealed, the son of perdition; who opposeth and exalteth himself above all that is called God, or that is worshipped; so that he as God sitteth in the temple of God, shewing himself that his is God. (2 Thessalonians 2,4)

Jester Crowns Himself Emperor

Letah and Verna Mae Stroh in Southern California

Chapter Five - Sisters in The City of Angels

IN 1859, ENGLISH author Charles Dickens published monthly installments of a forthcoming book in his London periodical *All the Year Round.* Titled, *A Tale of Two Cities*, Dickens' book has become the best-selling novel of all time. Set during the end of the eighteenth century between London and Paris, the historical novel describes the intrigue centered around fictional characters entangled in the French Reign of Terror (September 1793 to July 1794).

Considered to be the classic opening line for a novel, *A Tale of Two Cities* begins, "It was the best of times, it was the worst of times, it was the age of wisdom, it was the age of foolishness…—in short, the period was so far like the present period…"

Conflict and war create uncertain times. Yet anarchy with its inherent chaos terrorizes everyone. Totalitarianism is arbitrarily oppressive; anarchy is universally so. In Dickens' clash of contradictions, he took a page right out of Proverbs 9. There, wisdom and foolishness sit along the roadside mirroring similar enticements to the naïve passersby. The first woman calls as a wise muse; the other a foolish siren. Both women seek to draw in the reader who lacks understanding using a similar refrain:

> Whoso is simple, let him turn in hither: and as for him that wanteth understanding, she saith to him,... (Proverbs 9:4, and 9:16)

Wisdom as a Maiden with Bread and Wine

In Proverbs Chapter 9, both the wise woman and the foolish follow their respective invitations with contrasting enticements. Wisdom declares the reward of the two biblical elements of the blessing; something fat and something sweet. The bread and wine are openly available, offered with a cautionary admonition.

> Come, eat of my bread, and drink of the wine which I have mingled. Forsake the foolish, and live; and go in the way of understanding. (Proverbs 9:5)

Folly as a Seductress Before a Secretive Tent Door

By contrast, foolishness plays a shell game of deceit: a prisoner's banquet of bread and water hidden out of view.

> Stolen waters are sweet, and bread eaten in secret is pleasant. (Proverbs 9:17)

The Family Library and Music Room

As a teenager, I often wandered into the music and library den of our home to sift through the extensive collection of books; at times it was my mother's form of reprimand, "Go sit down and read a book." A glass-enclosed bookcase filled a section on the north wall of the room. One book caught my eye; an old illustrated edition of Charles Dickens, *A Tale of Two Cities.*

After a cursory skim through the book, I found it not to my interest and returned the volume to the shelf. Only later did I discover the troubling historical backdrop and literary significance of the novel.

Hollywood Rags

My mother Letah occasionally related stories of her college days in Los Angeles. After high school she attended Compton Junior College and UCLA (University of California at Los Angeles). It was there she met my father William McGreehan Sr. in 1945. What would attract a teenager in 1939 to leave Upstate New York for college in Southern California?

Louis and Hanna Stroh and Family (1923)

By the late-1930s, the nation and world gradually emerged from the Great Depression (1929-1939). My grandparents, Albert and Ruth Spring Stroh managed to build a very successful floral greenhouse operation on River Street in Batavia, New York. Their extensive greenhouses served as part of a floral business founded by his parents, Louis and Hanna Rosenkranz Stroh.

In the mid-1800s, Louis's parents William and Catherine Dietch Stroh emigrated from Wurttemberg, Germany. At that time Germany experienced much turmoil, as in most European regions when governments transitioned from post-Napoleonic city-states and reorganized as nation-states. America presented the Stroh family with an escape from the chaos of conflict and oppressive taxation.

Bread Line in Post World War I Germany

Several close relatives from Germany visited my grandparents Albert and Ruth Stroh in the mid-1930s, declaring how their country had quickly recovered from the hyperinflation of post-World War I Europe. Suddenly there were plenty of jobs available. Still, the specter of renewed conflict in Europe remained. My mother expressed how her family in Upstate New York continued to be very skeptical of events unfolding in her great-grandparent's homeland of Germany.

About this time my mother's next older sister, Verna Mae Stroh made a decision to apply for college at the newly expanded campus of UCLA in Los Angeles (City of Angels). After graduation from high school in 1939, my mother decided to join her sister and applied to UCLA where she studied music. She later transferred to Compton College in Los Angeles. Both she and her sister Verna Mae continued living in Southern California during World War II until the mid-1940s. Through music lessons and child-sitting, my mother became familiar with several people associated with the entertainment business in Hollywood.

Jester at the Box Office

Thomas Edison's early experiments in the 1880s-1890s with Vitascope video recording followed his successful development of phonograph cylinders. At the turn of the twentieth century live theater in the form of traveling Vaudeville acts experimented with the use of movie shorts during intermissions. These movies featured improvements to Thomas Edison's Vitascope filming process.

Beginning in the late 1920s with the introduction of movie sound, Hollywood entered its Golden Era. Suddenly theaters became an attractive and inexpensive distraction from the dark days of the depression.

As silent movies replaced live theater productions, most filming moved from New York to California where movies could be made year-round. Thus began the movie and entertainment industry centered in what is now the Hollywood section of Los Angeles. On the heels of this industry, magazine publications termed "Hollywood Rags" fed the public appetite for information about movie stars. My mother often rehearsed stories she read as a teenager about the cinema business.

Coal Miners of Appalachia in the 1930s

The Coal Mines of Appalachia

At the same time during the mid-1930s, my father's family in western Pennsylvania sought escape in the movies. In the coal mining region of Appalachia, the mining industry prospered despite the depression. Like most of the young men, my father quit school after the eighth grade to work in the mines, as much to meet the demand for coal as for financial reasons. Inexpensive movie tickets of twenty-five cents for adults and ten cents for children offered a diversion from life in the coal mines where working conditions were dirty and dangerous.

As mines opened and closed through the early 1900s, my great-grandparents and grandparents' families moved on average every two years. Sometimes a mine would shut down as the coal ran out; other times it would be due to a serious mine accident. Coal dust, methane gas, flooding, and mine collapse posed daily threats and risked a shortened lifespan.

Despite the challenges, my father always managed to work and save money, paying for everything including automobiles with cash. As an avid musician and classical music fan, he learned to play ten instruments, even purchasing a baby grand piano at one point. At the same time, he took on numerous creative projects such as making his own crystal radio set, doing photography development, 16 mm films, as well as building a telescope for himself.

Radio Announcement of Pearl Harbor Attack in 1941

World War II

While driving into Johnstown, Pennsylvania on Frankstown Road, the world suddenly changed for Dad. The radio announcer broke in with the news on December 7, 1941 of the surprise attack on Pearl Harbor. As America entered World War II, he volunteered to join the United States Army at age twenty-five. After working for a time in Butte, Montana mining copper to support the munitions industry, Dad got orders to ship to Hawaii and then to the South Pacific. There he served as a 40 mm anti-aircraft gunner on-board ship until he landed and stationed on Guam for most of the remainder of the war. In the Army, he volunteered to play clarinet in the marching band for his regiment.

Dad's Army Portrait in WW-II

By mid-war, the Army Air Corp (now the Air Force) advertised for desperately needed aircraft pilots among the troops and general public. Key requirements to qualify for pilot eligibility were 20-20 vision and a college education. To meet recruitment goals after Pearl Harbor, the Air Corp lowered the educational requirement for pilot training to least a high school education, or passing an equivalent intelligence test. After acceptance, candidates would be required to pass the Officer Qualification Test (OQT) with high enough scores to begin training.

While in the Army, Dad went to the recruiter's office for pilot training and asked to take the OQT, believing his technical aptitude would enable him to qualify. Instead, the recruiter immediately ushered him to the door; a big disappointment. It was not until after the war that he acquired his high school diploma and college courses through the GI bill.

Troops Returning from the South Pacific After World War II

That's Entertainment

By 1945, at the close of World War II in the Pacific, Dad returned to the West Coast, arriving at San Diego. Then, while awaiting transport by train back to Western Pennsylvania, Dad spent time in Los Angeles where he was invited to take an afternoon horseback ride. That is where he met Letah Pearle Stroh and they became acquainted.

Los Angeles, The City of Angels

Mutual interests in music, dancing, and the entertainment industry created an immediate connection. During the time they dated in the fall of 1945, Dad and Mom had opportunity to meet and socialize with some of the people and culture at the perimeter of the movie industry.

Over the years since then, he related how his perception of Hollywood and the entertainment culture changed dramatically. By the 1970s California and particularly Los Angeles acquired the nickname La La Land; a dreamlike existence out of touch with reality, absorbed by glamor, music, and movie making.

The Bizarre World of Hollywood

With the 1950s advent of television, the secret lifestyles of rich and glamorous stars went on full display. Still, that did not dimmish Dad's fondness for music, dancing, and the theater. In order to view his favorite variety shows, Dad grudgingly purchased a black and white television set, which he addressed as the "One-Eyed Monster".

Dad would often punctuate the program with comments about the personal lives of entertainers who did not live up to their public image. A favorite in our home in the 1950s and 1960s was the Lawrence Welk variety program which often featured singer Norma Zimmer.

Hollywood and the Entertainment Industry

During that period Norma Zimmer (1923-2011) occasionally sang at revival crusades of Evangelist Billy Graham, sharing a personal testimony of her faith. It was not until the publication of her autobiography *Norma* (1972) that the difficult side of Norma Zimmer's upbringing and personal life became public.

When I picked up a second-hand copy of the book years later, reading the parallels with our own family were striking. Though Norma's childhood during the Great Depression (1929-1939) presented typical financial and personal challenges, a specific event colored their family dynamic for the rest of her life. Details of that encounter as it relates to the theme of this book are mentioned in Chapter 12.

Jester in a TV Set

The Emerald City

The movie plot of *The Wizard of Oz* (1939) opened to an average day with Dorothy and her dog Toto on the family farm in Kansas. Three farm hands enter the scene in the midst of a tense confrontation with neighbor Almira Gulch who threatened to have Toto destroyed. In order to save Toto from Almira, Dorothy chose to run away with Toto. There along the road in an isolated vale, Dorothy came across a gypsy wagon of traveling salesman and fortune teller, Professor Marvel.

Upon consulting his crystal ball, Professor Marvel counselled Dorothy to return home because her Aunt Em looked worried sick. As Dorothy returned, the farm faced a looming tornado. Suddenly, the house with Dorothy and Toto inside became caught up in the whirlwind and transported to the dreamlike Land of Oz. We're not in Kansas anymore.

Through most of the movie, Dorothy, Toto, and a whimsical trio of a scarecrow, tin man, and cowardly lion weave along a yellow brick road toward the Emerald City, home to the Wizard of Oz. As they traveled seeking to have their requests answered by Oz, the Wicked

Witch of the West plotted their demise at every turn, even dispatching a flock of flying monkeys.

Despite plans of their adversary, the travelers eventually reached the Emerald City where Toto pulled back a curtain to unveil the "wizard" manipulating the levers and voice of Oz. After each of Dorothy's three companions received their request, she asked to return to Kansas. Glinda the "good witch" informed Dorothy to merely click the heels of her ruby slippers three times while repeating "There's no place like home". Her wish was granted and she returned to Kansas and her family.

Jester Behind the Curtain

While Debby and I discussed the movie storyline with a friend, she expressed surprise as I explained how each of the characters in Kansas had their expression as an alter ego in the Land of Oz. Despite watching the movie many times, she never realized how the characters and their counterparts in Oz combined to develop the mystery underlying the story.

The final chapter of this book frames the conclusion of the gospel message as the final act of a play *L'Acedia, Land of Leisure*. Everything is not as it seems. The characters each find themselves engaged in their pursuits, while behind the curtain there are influences at play to execute a plan. Others provide only resistance.

Too often entertainment distracts the audience from understanding the message. The same scenario plays out with the gospel message and the closing act at the return of Jesus Christ. Every story and historical scene fits into that overall theme in the closing chapter of this book.

Wright Brothers Flyer

Chapter Six - The Wright Stuff

FOR THE SAME reasons the film industry located in Southern California in the 1930s, aviation companies flourished in the Los Angeles area. Major airplane manufacturers such as McDonnell Aircraft, Douglas Aircraft, Glenn Martin Company, and Lockheed Aircraft had their beginning in California where favorable weather permitted year-round operation and flight. Test centers favored the California high deserts east of the Sierra Nevada Mountain range.

What relation does the aviation industry have to the theater stage and the theme of this book? Flying as a business requires everything be done right. It is unforgiving of error, unlike Hollywood. The ability to fly a heavier-than-air vehicle challenged adventurers and inventors for centuries.

Orville and Wilbur Wright are generally credited with the first successful powered flight, though there were other contenders for that title. The Wright Brothers achieved that flight in 1903 at Kitty Hawk, South Carolina with a specially designed glider and custom developed gasoline engine. A great amount of scientific investigation by the brothers preceded that original flight. Why did the Wright Brothers succeed where others failed?

Early Attempts at Flight

Crash Tests

The tour guide at one aviation museum prefaced his talk about the history of flight with the claim, "Before the Wright Brothers, no other experimenters did anything fundamentally right." All prior efforts involved either trial-and-error, or simply copying what they observed in nature. Whether by duplicating the use of feathers or the creation of flapping wings, those attempts failed because they did not seek to understand how powered flight actually worked.

The Wright Brothers however did something unique to achieve a flying machine. Working from their bicycle shop in Dayton, Ohio, the brothers investigated what they observed in nature to understand the mechanical principles involved. From there they determined what needed to be done to reproduce the same result at the scale of their proposed machine.

Specifically, that involved freezing dead pigeons and using a bandsaw to cut the wing. By duplicating the shape of the wing cross-section, they were able to measure the wing lift and drag using their own crude wind tunnel. Varying the air speed and angle-of-attack to wind direction, the Wright Brothers measured the lift and drag for a given size of their model "airplane wing". From those results they were able to calculate the engine size and power needed to sustain flight at a given speed.

CFM56-3 Jet Engine and 737-300 Aircraft

While that may seem to be a simple process, it required roughly four years for Orville and Wilbur to complete a functional glider and create a lightweight motor powerful enough to propel the machine to the necessary speed. In addition, that included developing a pair of chain-driven propellers, controls, and landing skids. On December 17, 1903 at Kitty Hawk, South Carolina, the Wright Brothers achieved several sustained flights, the longest of which was 59 seconds.

At that time, the automobile industry still remained in its infancy. Over the next thirty years aviation technology developed into a major factor in transportation, travel, entertainment, and warfare.

Today, the commercial and military aviation industry influences every facet of life. Having worked in the aviation and space propulsion field for over fifty years, the development process for complex flight systems is much more demanding than the early effort of the Wright Brothers. My business leader in the development of aviation gas turbines once described a modern jet engine as a symphony in metal; a fitting depiction.

Whether flying a plane, public speaking, or writing, the take-off is the easy part; executing a landing takes real skill and preparation.

Hypersonic Aircraft Demonstrator

As I explained in the recent book *Rend the Heavens ...and Come Down*, the 2022 Hollywood movie *Top Gun: Maverick* opened with a scene portraying a supposed Mach 10 hypersonic plane. Filming employed a physical mockup and computer-generated imagery of the hypersonic aircraft rather than the real thing. A few years prior I had the privilege of doing systems integration engineering on the then-secret dual-mode scramjet engine for the real hypersonic propulsion system; not the mockup. The project has since been de-classified and its existence released publicly.

Much like Hollywood, *Top Gun: Maverick* deals with fantasy covered by a veneer of reality. If the movie simply reflected the daily routines of aviation, it would contain all the excitement of a bus commute to work. The performing arts specialize in entertainment, excitement, and distraction from reality. That is the attraction of the theater. By contrast, the ultimate monotony would be observing someone going through life's mundane tasks: an audience watching a spectator watching spectators.

Jester Watching Man on Stage Watching TV

Rather than become overly philosophical about art and reality, it suffices for now to observe that even the words and images in this book attempt to represent reality. Whether words, photographs, drawings, or computer-generated art, they are only a feeble reflection of the real objects and events themselves.

For several decades while developing jet engines, my organization at the company had to submit a yearly IR&D (Independent Research and Development) report to the government. During that period, I managed several IR&D funded analysis and test projects supporting new engine development. Responsibilities also included preparing our group's section as well as the larger business IR&D report.

In the process of writing the yearly report, I was always fascinated to look back over each investigation and test to see how the project contributed to engine design objectives. One of the purposes of the report included justification of costs based on research results.

Successful or not, each step helped to set the stage for the completion of a functioning system or engine. Likewise, accounts in this book; historical, biblical, and personal, represent more than stories in their own right, but pictures that set the stage for something more critical.

Mount Rainer, Washington

The Tyranny of Unintended Consequences

While a student at Penn State University in early 1968, I made applications for summer employment. One of the positions involved a US Forest Service construction support role in Washington State. An acceptance letter came in the mail a few weeks later with an offer and assignment details. Location of the position was at the Forest Service Ranger Station in Darrington, Washington, northeast of Seattle.

Darrington resides adjacent the Sauk River valley at the western foothills of the Cascade Mountains.

Responsibilities involved staking out mountain logging roads for timber harvesting in the Mount Baker-Snoqualmie National Forest. A side benefit of the job was the opportunity to do mountain hiking and photography on weekends.

Darrington Ranger Station and Whitehorse Mountain

Not only was the job fulfilling, but also a challenging life experience. Thirty years prior, in the same area, singer Norma Zimmer was raised and spent time skiing before moving to the vicinity of Los Angeles. Norma's personal struggles feature prominently in her book *Norma* (1972) and mirror my own family's difficulties.

Upon exiting the flight at Seattle, the fragrance of pine and sight of mountain vistas overwhelmed the senses. The bus ride to Darrington was magical. Every snow-capped mountain seemed a new experience. After asking a local co-worker several names of mountains, he responded,

"You must be a stranger to Washington. They're just mountains. No one around here cares about their names."

Mount Baker, Washington

The summer intern job included bunk house provision at the Darrington Ranger Station compound. About a dozen college students bunked at the auxiliary building while working summer jobs. Every Saturday evening members of the local Baptist church stopped by to invite us to church service and dinner. One of the families would pick us up for Sunday service and afterward treat us to a delicious fried chicken dinner. It was my first experience in a Baptist church, and though the sermon message about knowing Christ personally did not immediately resonate, two years later the seed took root and sprouted.

The trail to Copper Lake became our most challenging weekend hike. Situated at over 5,000 feet in the Central Cascades, the trail traversed a wicked bog with biting flies that attacked in swarms. After rising above the tree line, a punishing climb replaced the biting flies. My co-worker Steve and I hiked every other weekend, each carrying thirty-pound backpacks which made the muscles burn as we climbed to the ridge line.

After reaching Copper Mountain Lookout Station we took a needed break. From there the trail led down to Copper Lake with spectacular views of the surrounding Mount Challenger, Mount Fury, and Bear Mountain. Though it was early August, Copper Lake still had a significant amount of ice and snow covering the lake.

By evening Steve and I descended to Copper Lake and setup camp. Trout in the lake began feeding so I pulled out a fishing line and gathered some live bait. By the time they stopped feeding, I managed to catch a nice evening meal of three trout.

After a short internal debate the next morning, I dove into the partially frozen lake and swam out to an ice floe and back. Occasionally after working on mountain roads the crew would stop at a river and cool off in near-freezing water. So, I was accustomed to a very cold swim. The lake was invigorating and the swim quick enough that my muscles did not cramp in the cold water.

Swimming in Copper Lake, Washington

Though the mountain views climbing the Cascades resulted in spectacular scenery, the strenuous trails gave little time for anything but occasional breaks and photo shoots. On the way downhill it was much easier to appreciate the vistas. At times the breathtaking panoramas defied description. While visiting the local Baptist church, workers in the timber products industry expressed how outside influences were forcing the Forest Service to close much of the national forest to the logging industry.

My travels during a fifty-year career included the most interesting cities and picturesque locations around the world. Increasingly, many of them have fallen under socialist control. In the book *Norma*, Norma Zimmer described her teen years around Seattle-Tacoma in the 1930s. As Norma related, Communist influencers moved into the Northwest to wrest control of labor and government.

During that period her father Peter Larson experienced a job layoff resulting from labor unrest in the area. When Norma Zimmer moved to Los Angeles in 1941 in pursuit of a singing career, she encountered a culture largely controlled by socialists in the film and entertainment industry.

In 1995, a privately funded project installed a statue of Vladimir Lenin on private property in the Fremont neighborhood of Seattle, Washington. The statue of Lenin had been rescued from a scrapyard in Czechoslovakia after the fall of communism in 1989. By 2020, socialist and communist-leaning leaders gained control of most major cities on the west coast.

Statue of Vladimir Lenin in Seattle, Washington

Be Careful What You Demand

One caution spoken by Moses related to the choice and nature of a king (Deuteronomy 17:14-20). Over two hundred years later, the people, under Samuel as judge, demanded a king like every nation. This deeply troubled Samuel, yet God authorized him to give the people what they demanded while admonishing them the consequence of autocratic rule rather than the judicial theocracy (1 Samuel 8:1-22) under God's rule.

In response to the demands of the people, God led Saul the son of Kish to Samuel who anointed him king and revealed Saul to the people. For a time, the kingdom functioned well. Saul and the people even defeated the Ammonites which helped cement his rule. Then the kingdom began a slide toward cautioned authoritarianism under Saul, disobedience regarding destruction of the Amalekites (Hebrew

– *Amal* meaning work or toil), and a breech with Samuel. In turn, God directed Samuel to anoint young David (son of Jesse) as king instead.

Principles in Scripture can be understood by the meaning of associated Hebrew words. The name of Saul is no exception. When the people demanded a king, God gave them a man named Saul (meaning to ask for). The name Saul and Sheol (hell or underworld) both have origin in the Hebrew word *shaal* or demanded.

The concept of Sheol reflects the consequence of bad choices. Some ask critically how a loving God can send people to hell. That question, usually framed as a challenging accusation, seems an irresolvable conflict to the nature of God. The biblical concept of Hell echoes the warning that parents give concerning the local bar, "You don't want to go there."

Hell is a place you don't want to go, not a threat. It is a place of separation from God; don't go there. Disobedience and evil cannot exist in the presence of God; therefore, He created a place of separation, or Sheol, meaning that's what you demanded. Insisting on your own way or something other than the plan and purpose of God can mean He gives you what you want, saying, "That's what you demanded."

Thank God there is coming a rest when God's purpose on earth will be complete. That time will be a personal rest for believers as well as rest for the earth and a government where believers rule and reign with Christ for a thousand years (Revelation 20:4-7), and finally spend eternity in glory.

The principle of what you demanded has implications to career choices, relationships, even the outcome of elections. Increasingly voters select leaders, like Saul, who turn their back on God. Samuel rehearsed the consequence to the people of Israel, who replied, "Nay; but we will have a king over us…" (1 Samuel 8:19). The name Saul vividly illustrates the outcome. Be careful what you demand.

Cycle of Self-destruction

Human nature strikes again. Moses warned the people in the wilderness of their inevitable falling away. In the list of blessing and cursing (Deuteronomy 28), a short list of blessings for righteousness precedes an extensive list of curses for disobedience.

With their insistence of a king, each successive leader began a swing toward greater evil until a righteous king arose. The cycle then repeated itself with more extreme evil and a greater turning back to God. Over a few generations the oscillations between evil and righteousness became more extreme. Finally, God pronounced judgment and Babylon conquered Israel. The cycle again started over.

Greek philosopher Polybius termed the degenerating leadership cycle *anacyclosis* or "again-turning back". Polybius claimed the inevitable decay reflects human nature and corruption. He listed the stages as: Monarchy, Tyranny, Aristocracy, Oligarchy, Democracy, Anarchy. And with anarchy comes self-destruction because chaos oppresses everyone equally.

The Oppressed Becomes Oppressor

English author George Orwell (1903-1950) wrote two popular books exposing the abuse of power in authoritarian government; *Animal Farm* (1945), an allegory of the underclass of pigs turned oppressors, and *Nineteen Eighty-Four* (1949), a story of government control devolved into a dystopian nightmare. Both books emerged out of popular uprisings gone off the rails. Born Eric Arthur Blair in India, he wrote under the pen name George Orwell.

Though he regarded himself as a democratic socialist, Orwell witnessed the slide toward authoritarianism in the Marxist movement of Russia turned Soviet totalitarian as well as national socialism in Germany turned fascism. Both socio-political uprisings began as working-class movements against poverty and the establishment.

Quickly they evolved into authoritarian dictatorships and then destruction. Orwell remained a socialist, despite both books warning of the inevitable slide of socialism toward repression. A society in the process of self-inflicted decline is often referred to as "Orwellian".

A Pork Barrel Experiment

Fifty years ago, my home district of Southwestern Pennsylvania suffered from a serious decline in the coal and steel industries. Coal mining and steel production, the economic foundation of the region for the past century, put up a "For Sale" sign. Though decay began decades earlier, the so-called Western Pennsylvania Rust Belt turned into a virtual welfare state.

For most of his life, my father worked in either the coal or steel industry. While out of work for several years he took odd jobs, refusing to accept welfare relief. But as the crisis deepened, Dad like many others, finally signed up for benefits. Government assistance and surplus food provided a lifeline for our family during that time.

At the peak of the economic downturn, the newly elected federal congressman for the region began to promote what are called "pork barrel" projects. Pork barrel refers to programs introduced by a member of congress with funding directed to the home district. Often those projects involved "no bid" contracts to local businesses setup solely for the purpose of administering a funded government program.

Of course, no "pork" project would be complete without a generous donation back to the sponsoring member of congress. Though often referred to as "welfare manufacturing", the process provided some economic benefit, especially to the politician, family, and friends. Even my father felt obligated to continue voting for the status quo, despite complaining furiously about the corruption and waste.

In this instance, the congressman came under federal investigation for corruption, but upon his untimely death the probe was dropped. Warnings of Moses, Polybius, and Orwell each have a chilling premonition as the consequence of human self-interest.

Ben Adhem

During the nineteenth-century, European culture developed a fascination with the Orient. In 1834, writer James Henry Leigh Hunt (1784-1859) penned a poem titled *Abou Ben Adhem.* The subject of the poem, Ben Adhem, was thought to be a devout Muslim Sufi nomad named Ibrahim ibn Adham who abdicated his throne and wealth to be a wandering cleric. The name *Abou Ben Adhem* in Arabic means: father, son of man. Hebrew and Arabic names are both of semitic origin and similar in appearance and sound.

Hunt's poem begins with Ben Adhem visited in a dream by an angel making a list of names of "those who love the Lord". When Ben Adhem enquired whether his name was one of them, the angel replied, "Nay, not so". To that Ben Adhem said, "I pray thee, then, write me as one that loves his fellow-men."

Here the poem slides into virtue-signaling humanism. The next night the angel reappeared revealing a list of those God had blessed; "And, lo! Ben Adhem's name led all the rest." The Kingdom of God builds on faith in the completed work of Jesus Christ. Love of fellow-man (and enemies) is a fruit of relationship with God but of itself does not earn merit with God. Trying to assign reward to the works of man builds on the vanity project of Nimrod, the city of Babylon (meaning chaos).

The entertainment industry today continues to market delusion and distraction at the expense of truth. Meanwhile, the audience follows down a moral slippery slope. By contrast, the reward of faithful obedience and service will shortly become clear. Jesus Christ comes quickly to rend open the heavens beginning a final chapter of His kingdom.

[illegible] The name [illegible] in Arabic means [illegible] how [illegible] made often and [illegible] and found.

[illegible]

[illegible] God [illegible] to the works of man's hands [illegible]

Old Pickup Truck in a Rut

Chapter Seven - Don't Focus on the Ruts

FOR THE FIRST few months after moving to an apartment in Tulsa, Debby and I had only occasionally seen our neighbors below. We greeted the couple a few times though they seemed distracted. But we noticed a daily procession of visitors at their apartment, each of whom would arrive at their door and leave a few minutes later.

Most days pass with a simple, predictable routine. Those times would be difficult to imagine as entertaining theater. Occasionally drama interrupts the ordinary and suddenly a script worthy of Shakespeare unfolds.

It began with a simple request to the neighbors in the middle of the night. On several occasions the couple in the apartment building below ran a vacuum or turned up the television set between midnight and two in the morning. Floors and walls of the aged building in Tulsa, Oklahoma did not absorb much noise, and it sounded as though they intentionally created as much of a disturbance as possible after midnight. On some nights the sounds would wake one or both of our young daughters in their bedroom.

As the nightly disturbances continued, I made a point to go downstairs late one evening and politely ask that they turn down the television at night. The couple both came to the door and he very graciously apologized for the disturbance, agreeing to turn it down. That solved the problem; or so it seemed. Several days later the disturbances began again. In fact, it seemed the sounds were purposely louder. So, a second time I went downstairs and knocked, asking him politely to lower the sound. His response again, though abrupt, was in agreement.

From there it started slowly but escalated quickly. Someone used a broom handle at night to bang on their ceiling. Conversation between the couple became so loud as to be clearly audible. They made no secret their topic was directed toward us as the neighbors upstairs. If we spoke to each other, they spoke back in obvious jest. Not even a cough went unanswered without a loud series of coughs downstairs.

One afternoon while I was working, Debby and the girls had another neighbor's daughter over for a visit. As the girls played there was a knock on the door. Debby answered the door, surprised by a police officer asking about the party inside. She assured him the girls were only playing in the apartment. So, Debby explained to him the previous interactions with the neighbor downstairs, reasoning that their complaint likely involved retaliation for my requests about their noise.

At that point the nighttime disturbances escalated further. This time I called the police and asked that an officer come to our apartment first so we could explain the background. When the officer arrived, he stated that he wanted to speak to the couple privately before bringing me into the conversation. After a short discussion downstairs, he came back and asked me to join them. I explained to our neighbors that we were only in Tulsa for a few more months on a temporary work assignment supporting the American Airlines jet engine maintenance center. During that time, I requested there be no disturbance or conflict.

Probation and Parole

You Know How It Happens

The policeman then addressed the couple explaining how these situations often spiral out of control. He said words and actions escalate until somebody pulls out a gun and the other is killed. The officer then addressed the man specifically saying, "And you know personally how it happens." To that the man nodded his head and responded, "Yeah, I know." The implication struck me immediately that the officer knew the neighbor's history.

Only a few days later, as I was gathering our mail from the mail rack in the lobby, I noticed a letter sticking out of our neighbor's mailbox slot for the mailman to pick up. Clearly printed on the letter's address were the words "Tulsa Probation and Parole Department". The letter confirmed my suspicion: he knew exactly how escalation happens. It occurs because, when you focus on the ruts, you always find yourself back in the ruts.

If you drive an old pickup truck through a field, it eventually creates a series of ruts through the mud. Always taking the same path forces the driver to focus on staying out of the ruts. When focusing on the ruts, you end up back in them. Avoiding the problem requires taking a completely different path.

While doing church jail ministry for about twelve years in Butler County, Ohio, the same pattern often repeated with inmates. Someone has their life turned around through the jail ministry and they are finally released. Going back to the same "field", and following the same "road", while trying to stay out of the "ruts", invariably you end up back in the ruts. For example, the person who declares, "Every day I tell myself to stay out of the bar", will eventually end up back in the bar.

When Sailing, Focus on Your Objective

Focus on the Objective, Not Obstacles

When sailing, there are so many influences happening at once, it is easy to lose sight of the objective. Focus on the objective and not on the obstacles. Aircraft pilots refer to that as "situational awareness": when flying, don't let any distraction cause you to forget that returning to the ground safely is your primary objective.

Breaking the cycle requires focus on a completely different path. It takes finding a purpose beyond just, "staying out of the ruts". I have seen many in a similar trap, and they succeeded only because they found a new path and purpose, by ministering to others rather than focusing on, "staying out of the ruts".

In Chapter 12 the history and role of the jester are explained. Whether the royal court or entertainment or games, the joker disrupts the predictable with a distraction to the script.

Tertullian - Christians to the Lion

Early church father Tertullian (about 160-220 CE) wrote extensively to document the basis for Christian faith. Born in Carthage, Africa as Quintus Septimius Florens Tertullianus, he produced a wide-ranging output of literature in Latin to support Christian teachings. One of the themes by Tertullian regarded the importance of persecution and martyrdom which he claimed fosters growth of the church.

Christians in the Arena Facing Martyrdom

In Tertullian's *Apology* he argued against Roman accusations of Christian treason and immorality stemming from their refusal to worship Caesar and the Roman pantheon of gods. There he declared, "the blood of the martyrs is the seed of the church". Commenting on the trend for public officials to blame every disaster on the Christians for a lack respect to the pagan gods, Tertullian wrote,

> If the Tiber rises as high as the city walls, if the Nile does not send its waters up over the fields, if the heavens give no rain, if there is an earthquake, if there is famine or pestilence, straightway the cry is, 'Away with the Christians to the lion!' What, all of them to one lion?

Tertullian also encouraged Christians in his writing *To the Martyrs* not to flee or bribe judges to escape martyrdom, but to view it as a victory and the ultimate reward. Beginning with the stoning of Stephen (Acts 7:58-60), Christians have faced various periods of selective persecution and occasionally severe tribulation, including martyrdom. In fact, the Greek word *martus* translated as witness also means martyr. Before His ascension Jesus declared,

> But ye shall receive power, after that the Holy Ghost is come upon you: and ye shall be witnesses [or martyrs] unto me both in Jerusalem, and in all Judaea, and in Samaria, and unto the uttermost part of the earth. (Acts 1:8)

The Jester as Theater Critic

A recent book, *The Myth of Persecution, How Early Christians Invented a Story of Martyrdom* (2013) by Candida Moss, seeks to minimize Christian tribulation and martyrdom as largely fictional. Moss, in her poorly substantiated narrative argues against early Christians experiencing persecution and death for their faith. She bases her premise on the contention that oppressive groups always fabricate stories to make themselves appear the victim.

Candida Moss, currently in the position of Edward Cadbury Professor of Theology at the University of Birmingham, launched her leaky-boat "myth" theory claiming that Jesus went to the cross disillusioned and defeated. The early church, in Moss's view, tried to rescue the crucification narrative based on pagan traditions by portraying death as noble rather than as defeat. Many skeptics of New Testament history rushed to aid her floundering boat. But eight chapters later Moss's boat sinks; not surprising, there were no survivors.

Perpetua and Felicity

One of the best documented events of martyrdom occurred the year 203 CE in the Roman province of Carthage where Tertullian lived. The death of the upper-class woman Perpetua and her servant Felicity is an account of two Christian women based largely on the prison diary of Perpetua herself. An observer, Saturus, wrote the introduction and conclusion to her diary account.

The Execution of Perpetua and Felicity

Perpetua, a recent mother at age 22, and Felicity were arrested during a Christian gathering. They both received a death sentence subject to the decree of Roman Emperor Septimius Severus. Despite appeals from friends and family to renounce their beliefs, they chose instead to go to the games in the Roman arena of Carthage with other Christians to be executed before the crowds. When both women survived the attack of bulls, an inexperienced gladiator received the call to behead them in the arena. After stabbing Perpetua, the gladiator hesitated whereupon she took his hand and steadied the sword against her own neck so that he could complete the execution.

Of twelve original apostles, Judas took his own life, and John survived an attempt to boil him in oil, later dying in his 90s of old age. All ten of the other apostles died by martyrdom. Death sentences and execution of first century Christians were pursued for a variety of reasons under Emperor Nero (54-68 CE).

It was during Nero's reign that the Apostle Paul was beheaded after his second imprisonment in Rome. Historian Tacitus wrote in *Annals*, how Nero deflected suspicion from himself for the burning of Rome by condemning Christians for the deed. Tacitus demonstrates no sympathy to martyred Christians.

> To kill the rumors, Nero charged and tortured some people hated for their evil practices—the group popularly known as 'Christians'. ...In their deaths they were made a mockery. They were covered in the skins of wild animals, torn to death by dogs, crucified or set on fire—so that when darkness fell they burned like torches in the night. Nero opened up his own gardens for this spectacle and gave a show in the arena, where he mixed with the crowd, or stood dressed as a charioteer on a chariot.

Systematic persecution did not start until the end of the first century under Emperor Domitian (81-96 CE) and continued with varying severity through the reign of Decius (249-251 CE). Emperor Decius issued an edict that all citizens make a token sacrifice to traditional Roman gods, whereupon Rome granted a certificate of compliance. Some Christians submitted to the Roman edict or obtained a certificate through a bribe. Those who refused and disobeyed received an order of public execution.

During this period, taking the label Christian became a crime. Because Christians refused to utter the epitaph "Caesar is Lord", they became suspect of treason. Refusal to acknowledge the Roman gods was believed to threaten peace and prosperity.

With the conversion of Emperor Constantine to Christianity in 312 CE, the Roman Empire quickly transformed from adversary to ally of the church. Although the alliance of church and state vacillated from cooperation to conflict, officially sanctioned martyrdom ceased until the Middle Ages. Lacking persecution, the church grew complacent and spiritual vitality waned.

Throughout the fifth to fifteenth centuries, state leaders and the officially sanctioned church applied the death penalty for those considered to be heretics. Occasionally accusations of heresy resulted in execution of church reformers like Jan Hus (burned in 1415) and William Tyndale (burned in 1536).

Telemachus Enters the Arena to Stop a Gladiator Fight

After the execution of Christians in the amphitheater ceased, gladiator games sanctioned by the Romans continued until about 404 CE. When a Christian monk named Telemachus entered the arena to interrupt a gladiator fight, the crowd became incensed. They rallied for the gladiators to kill Telemachus. Finally, a group of spectators descended to the arena and stoned Telemachus to death. The killing moved Emperor Honorius, himself a Christian, to ban violent coliseum games.

Hosea and The Valley of Achor

Life occasionally serves up a reality check. After successfully destroying Jericho upon entry into the land of Caanan, the people of Israel set their sights on a small, sparsely populated city of Ai (*Ai* – heap of ruins). A group of advance scouts observed Ai and returned to advise Joshua dismissively that he only need send a small band of warriors to overthrow Ai; a classic case of overconfidence (Joshua 7:1-9).

Before their attack on Jericho, God had warned Joshua that all the spoil of that city would be *cherem* (Hebrew – devoted or accursed). That meant all silver, gold, brass, and iron would be dedicated to the Lord and everything else appointed for destruction. The Hebrew word *cherem* implies it was to be set aside for no useful purpose. Unbeknown to everyone, Achan, son of Carmi of the tribe of Judah chose to take for himself an expensive Babylonian garment, two hundred shekels of silver, and a wedge of gold, which he hid under his tent.

Achan Burying Spoils of Jericho

Immediately after approaching Ai, the three thousand attackers were chased off upon suffering thirty-six casualties. Joshua, in a state of disbelief, fell before God wondering why they were suddenly on the defensive. God responded to Joshua telling him to get up off his face and deal with sin in the camp. After identifying the culprit as Achan (*Achan* – meaning trouble), Joshua condemned him and all his possessions to be stoned and burned in the Valley of Achor (*Achor* – also meaning trouble) for troubling Israel. But that is not the end of the story.

In the book of Hosea (a form of the name *Yeshua*), God called the prophet to marry a harlot Gomer (Hosea 1:2,3) who proceeded to be unfaithful to him. Despite Gomer bearing several children outside of the marriage, God promised reconciliation. The imagery of the Book of Hosea portrays the nation of Israel as an unfaithful wife who continues to be given to spiritual adultery despite God's best effort at redemption. Referring to the illegitimate children, God said of Israel, "...ye are not my people, and I will not be your God." (Hosea 1:9)

Finally, God promised to redeem His people by drawing them into the wilderness where the Valley of Achor (trouble) would become a door of hope.

> Therefore, behold, I will allure her, and bring her into the wilderness, and speak comfortably unto her. And I will give her her vineyards from thence, and the valley of Achor for a door of hope: and she shall sing there, as in the days of her youth, and as in the day when she came up out of the land of Egypt. (Hosea 2:14,15)

That trouble took the form of Gomer eventually being sold as a slave. Rather than write off his wife Gomer, Hosea went to buy her back for the price of a slave (Hosea 3:1-5). The reconciliation of unfaithful Gomer to her husband Hosea represents God's redemptive effort on the cross to pay the price for anyone willing to accept His sacrifice.

Hosea Buying Back His Wife on the Slave Market

Jesus paid the full penalty for sin on the cross. In the end, the valley of trouble instead represents a door of hope. That promise applies to any individual now as well as to the remnant of the Jewish people in the latter days after Christ returns.

> Afterward shall the children of Israel return, and seek the Lord their God, and David their king; and shall fear the Lord and His goodness in the latter days. (Hosea 3:5)

Planned Disruption

When King Benhadad of Syria (Aram) boasted how he would destroy Samaria by turning it to dust (1 Kings 20:10,11), King Ahab of Israel replied, "Let not him that girdeth on his harness boast himself as he that putteth it off." Competitors try to intimidate one another by the psychology of "trash talk" or "talking smack".

To counter taunting, elite athletes often train with a coach using a technique called "planned disruption". The use of interference helps an athlete tune out chaos in order to calm emotions and focus on the task at hand. From Genesis (10:8-10) to Revelation (17:1-18), the chaos of Babylon plays a strategic purpose in God's plan to complete His kingdom.

Tribulation: A Door of Hope

Tribulation and death do not represent failure or defeat. In the final chapter of this book, the story unfolds as God Himself enters history again to complete His plan. This time God has ordained a timetable to be unveiled before the return of Jesus Christ. We do not know the hour or the day, yet events reveal the signs. Just as God revealed to Daniel the timing for the first appearance of His Son in the sixty-nine sevens (Daniel 9:25), He also makes known the times and seasons of the second coming during the final seven *(shabua)*.

> But of the times and the seasons, brethren, ye have no need that I write unto you. For yourselves know perfectly that the day of the Lord so cometh as a thief in the night. For when they shall say, Peace and safety; then sudden destruction cometh upon them, as travail upon a woman with child; and they shall not escape. For ye, brethren, are not in darkness, that that day should overtake you as a thief. (1 Thessalonians 5:1-4)

As in the beginning God said,

> ...Let there be lights in the firmament of the heaven to divide the day from the night; and let them be for signs, and for seasons, and for days, and years: And let them be for lights in the firmament of the heaven to give light upon the earth: and it was so. (Genesis 1:14,15)

The stage has been set for the unveiling of the events leading up to the final trump when the heavens rend open at the appearance of Jesus Christ (Revelation 6:12). To avoid falling into the ruts along the way, it is necessary to take another road.

Jester as Judge

Chapter Eight - Lest You Be Judged

AN OFT REPORTED story states that well-known Christian author G.K. Chesterton once wrote a short, cryptic reply to the London Times in response to a request for an article addressing the question, "What is Wrong with the World?" As related, Chesterton's early 1900s article submission stated merely, "Dear Sirs, *I am*. Sincerely, G.K. Chesterton".

While there is no documented support for the specific details, in the process of researching the story, it appears Chesterton did write a letter to the editor of the Daily News in London on August 16, 1905 where he declared, "The answer to the question, What is wrong?' Is, or should be, 'I am wrong'."

The fundamental truth of this anecdote rests on the awareness that human nature tends to see the problem out there, and the solution from myself. By contrast, my problem resides in me, and the solution needs to come from outside myself; that is, transcendently revealed.

In the book *Rend the Heavens*, I presented the scenario of M&Ms floating inside the International Space Station (ISS), asking why they appear to float rather than fall to the floor.

ISS and Peanut M&Ms

Almost everyone who responds to that question says, "There is no gravity". That is completely wrong. If there are no windows in the ISS module, it would be virtually impossible to sense the force of gravity.

The only reasonable way to determine there is something termed "gravity" would be if information comes through a window from outside. That information enlightens me that the ISS rotates around the earth every ninety minutes, meaning something tends to balance the force of gravity; in this case centripetal force of rotation.

Without going into a detailed technical explanation, the principle reveals I cannot make a right judgment without a reliable external reference frame. Whether asking about moral absolutes or the answer to Pilate's question, "What is truth?" (John 18:38), the response needs to come from some absolute.

First consider moral truth. Though every person senses some basis for right and wrong, a moral judgment is often challenged with, "Judge not" (Matthew 7:1), or an appeal to social norms. In Matthew, Jesus wasn't putting out a blanket warning against judging, but rather pointing out how we will be judged by the judgments we make.

In the broader sense of judgment, Jesus said,

> …and why even of yourselves judge ye not what is right? (Luke 12:57)

What is meant by judgment? The Greek New Testament has several words variously translated "to judge" or other derivatives of the word judge. There are six key Greek words translated as judge, all variants of the Greek root *krino*.

Investigate: *Ana-krino* (to interrogate or examine)

Discriminate: *Dia-krino* (to separate or make distinction)

Judge: *Krino* (to judge or select)

Sentence: *Epi-krino* (to decree or sentence)

Condemn: *Kata-krino* (to condemn or censure)

Hypocrite: *Hypo-krites* (an interpreter or actor)

Consider these words in the context of the system of justice. If someone reports a crime, the first step involves the gathering of evidence or an investigation (*Anakrino*) by an enforcement body such as the sheriff or police. The investigation process determines whether or not evidence of a crime exists.

As a result, evidence goes to a grand jury to compare against the law (*Diakrino*) and identify if there is justification to charge someone for the crime. If charges are filed, the accused goes before a trial hearing by jury or judge (*Krino*) in order to determine guilt or innocence. That process renders a decision of guilty or not guilty (*Epikrino*). In the case of a guilty verdict, a sentencing hearing is scheduled to pronounce punishment (*Katakrino*).

By judging one another, the individual takes the role of law enforcement, grand jury, judge or jury, and executioner. The consequence of usurping the judgment process makes the one who judges a hypocrite (*Hypokrites*—an actor playing a part). Therefore, Jesus said,

> Judge not, that ye be not judged. For with what judgment ye judge, ye shall be judged... (Matthew 7:1,2)

Every day the news is filled with examples of what was once evil; now championed. Paul summed up the matter of judgment in his first letter to the Corinthians:

> For if we would judge ourselves, we should not be judged. But when we are judged, we are chastened of the Lord, that we should not be condemned with the world. (1 Corinthians 11:31,32)

Actor Playing a Part

In Chapter 5, I related the story about my dad's experience in Los Angeles after World War II. Like many Americans during the post-depression era, he loved playing music, dancing, and the theater. The aura of Hollywood captured imaginations through radio dramatization as well as in the movie theater especially after the introduction of sound and color.

Suddenly, in Los Angeles he saw first-hand the reality of life in the entertainment culture. Hollywood represented something foreign to what he saw on the big screen. In the 1940s, show business still maintained a veneer of respectability in film themes as a result of government regulation and industry standards. Gradually the entertainment industry succumbed through the process of deregulation, pressure from special interest groups, the influence of film renegades, and the lure of financial gain.

Today the acting industry has become a display of the definition of hypocrisy; an "actor who plays a part". Actors who campaign against guns but brandish them before a camera. Or they speak out against violence yet act it out on film. And then they rightly excuse themselves, "We are just actors playing a part, or showing reality."

Does Hollywood change the culture, or vice versa? The reality involves something much more sinister.

The Pied Piper of Follywood

In 1842, Robert Browning wrote a fanciful poem, *The Pied Piper of Hamlin*, a moralizing tale in rhyme. The story tells of the citizens of Hamlin plagued by rats. A corrupt mayor, enticed by a colorfully dressed minstrel, accepted his offer to enchant the rats by hypnotizing them with his musical instrument. Of course, the agreement included a significant fee of a thousand guilders. The mayor and townspeople reluctantly agreed.

The Piper Leading Rats Out of Hamlin

After successfully charming the rats out of town and into a nearby river, the minstrel received instead only fifty guilders. The mayor and town council by now assumed they had the piper against a wall because the rats could not return.

But the minstrel then proceeded to play a tune charming all of the children of Hamlin. He led them out of town and through a portal, never to be seen again. In conclusion, Browning's poem advises that, agreements must be kept with all men, especially pipers.

Satan and his minions (a Jester) create nothing, but maintain the capacity to corrupt anything to their own purpose; thus the charm of Babylon. Satan acts as the salesman. Eve, for her part, accepted the serpent's subtle suggestion that God was holding out on them. Like the seductive siren (Proverbs 9:13-18), the serpent played to the lust of the flesh (good for food), the lust of the eye (pleasant to the eyes), and the pride of life (desired to make one wise).

In the garden Adam, Eve, as well as the serpent (Satan manifested) each had to "pay the piper" for their own part in the fall. As observed by G.K. Chesterton, the problem does not reside out there with the world (or Follywood); but "I am the problem".

Don't Fear Discipline—Fear Consequences

During His first year of ministry, Jesus departed from the region of Judea around the Jordan river and traveled to Galilee. While passing through Samaria, he stopped at the Well of Jacob near the city of Sychar (John 4:3-6). Situated between Mount Gerizim and Mount Ebal, Sychar corresponds to the ancient city of Shechem. There Jotham son of Gideon pronounced his parable against his half-brother Abimelech and the men of Shechem. The moral of that parable was, "Bad choices breed evil consequences". Abimelech's outcome fulfilled the prediction of Jotham. More about Jotham and Abimelech in Chapter 12.

Jesus with the Woman at the Well

At the well, the twelve disciples departed to buy food. As Jesus rested near the well, a Samaritan woman approached to draw water. She expressed surprise that Jesus asked her for water because historically the Jews had little interaction with Samaritans based on their corrupted beliefs and practices. This followed from ancient idol worship of Baal-berith at the time of Gideon. After the Jewish dispersion to Babylon the people of Samaria blended pagan practices with worship of Yahweh, and were referred to as Samaritans.

As Jesus shared with the Samaritan woman that He was the living water, He also disclosed that she had five husbands and was living with someone who was not her husband. The law (Leviticus 20:10) declared the death penalty and stoning specifically (Deuteronomy 22:22-24) for both parties of adultery.

In the conversation with the Samaritan woman as well as the woman accused of adultery (John 8:3-11), Jesus made no suggestion of the prescribed punishment. Jesus Himself is the Son of God and the Word of God specifically (John 1:1-4, 1 John 1:1,2). Therefore, Jesus is author of the penalty. Why does Old Testament Scripture not record instances of men and women being stoned for adultery? Does it seem a contradiction that Jesus made no suggestion of stoning to either of these women? And where were the men?

Regarding judgment, Jesus said to the paralytic after healing him at the Well of Bethesda, "...Behold, thou art made whole: sin no more, lest a worse thing come unto thee." (John 5:14). Jesus did not pronounce judgment for the paralytic's past but instead healed him. When accused by the Jews of healing on the Sabbath, Jesus responded, "For the Father judgeth no man, but hath committed all judgment unto the Son:..." (John 5:22). Yet, instead of coming to judge, Jesus came to pay the price for sin so that He could show mercy.

> But go ye and learn what that meaneth, I will have mercy, and not sacrifice: for I am not come to call the righteous, but sinners to repentance. (Matthew 9:13)

Scripture, particularly the book of Proverbs, employs rhythmic poetry in the form of comparing and contrasting principles. Actions and their respective consequences are equated to emphasize right choices. The word rod in Hebrew means correction as well as a physical instrument of discipline.

Discipline requires teaching children that choices have consequences. Fear consequences; not correction. Correction is necessary, but a child raised to fear only correction grows up making choices to escape correction. A child raised to respect consequences corrects their choices to achieve right consequences. Use correction, but teach consequences.

Jester Selling Stones

Planting Corn

Chapter Nine - Don't Eat Your Seed Corn

After attending the 50th Reunion of my Class of 1966 at Triangle Area High School, we took a side trip on the way back to Ohio. We had talked for several years about visiting Amish country in Holmes County, Ohio. The city of Berlin (BUR-lin), Ohio resides in the center of the Amish region and contains many shops and interesting sights.

By the time we arrived after sundown, the narrow roads through Amish farm country had become completely darkened as the Amish do not use outdoor lighting. We finally stopped at one of the beautiful Amish-built hotels in Berlin, hoping to find a room for the night.

Amish-Built Hotel Lobby

At the registration desk, when I asked about a room, the receptionist explained that in the fall season rooms are usually booked a year in advance. She apologized saying, they rarely have rooms available except due to infrequent cancellations. When I returned to the car and explained the situation to Debby, she said, "Go back in and ask them to call if there is a cancellation this evening."

Reluctantly I went back and gave the receptionist my number, just in case. By this time, it was already 8 pm, so I thought it would be highly unlikely. While standing at the counter, she exclaimed they just had a last-minute cancellation. She was as amazed as I.

Amish-Built Hotel Lounge

Amish Carpentry

After many years of doing finish carpentry and furniture as a hobby, I enjoyed the opportunity to see the fine Amish craftsmanship throughout the hotel woodwork. If you know anything about woodworkers, the first thing they do when exposed to finished wood is to run their hand over the surface to check fit and finish of the workmanship.

The next morning after breakfast at a large Amish run restaurant, we returned to the hotel to checkout. Debby asked if we could stay another night, and I agreed to ask, even though it seemed unlikely to experience another cancellation. But when I checked, the receptionist again said a room just became available. And it was the same room. We spent the rest of the day touring through the area and visiting shops and restaurants in Berlin.

While walking through an Amish owned furniture store, I spoke to one of the sales persons asking where the furniture was manufactured. She said the factory located across the road made most of the furniture, and if I wanted, she would ask the shop manager to give me a tour.

So, we arranged a time and I took Debby back to the hotel before returning for a visit to the factory.

Amish Farmers Doing Woodworking in Their Barn

The factory manager explained prior to the tour that Amish farms in the area make virtually all of the furniture components. He said, "This shop only does final assembly and finishing." Many of the local Amish farmers and their families produce specific parts like drawers and doors in small shops on the farm. That supplies extra work and income during free time and the winter season. He noted, Holmes County has close to 100 percent employment, one of the highest in the state of Ohio.

That result comes from the Amish philosophy of distributing work to keep every person gainfully employed. Unlike a socialist culture that distributes the fruits of labor to create parity, the Amish distribute the work so everyone contributes something. The factory manager said the Amish owner insists, where ever possible, spread work out to where the people live rather than forcing laborers to commute to a factory. Particularly where most Amish travel by buggy, bicycle, or rental car, that creates a great convenience.

As a beginning engineer working in a small equipment manufacturing business, we regularly made decisions about whether to fabricate a part or subcontract it to a supplier. The choice depended on the cost, complexity, timing, and sustainability.

Historically, many businesses have attempted to control as much of their supply chain as possible rather than just concentrating on their core skill. At the time I was doing design, drafting, marketing, and some fabrication. Out of necessity to complete an expedited equipment order, I often went into the shop and manufactured parts myself.

The alternative involved purchasing off-the-shelf components or, if essential, negotiating with multiple suppliers to make the part. The owner of that business would reflect on an old farming adage, "Don't eat your seed corn." Thinking ahead by developing partners and strategic alliances in business is an important part of long-term success. Often success breeds complacency; doing things the way they have always been done.

The Book of Job offers a view of man's life with everything taken away and only his wife left to counsel him, "Curse God, and die." (Job 2:9). Faced with three friends trying to comfort him by finding his hidden transgression, Job set out on a journey of self-justification. Little did Job know he had an adversary; a jester behind the curtain working against him.

Not until well into Job's defense of himself did a fourth friend Elihu (Hebrew meaning his God) reveal himself. Over the next six chapters (Job 32-37), Elihu stood as advocate for Job by correcting both he and his three friends for their respective misunderstanding.

Elihu with Job and His Three Friends

In the book *Rend the Heavens*, I related our childhood family story of a similar series of personal tragedies and setbacks. In that environment, seeing the purpose behind it requires faith and a vision through the veil to God's bigger plan.

> And we know that all things work together for good to them that love God, to them who are the called according to his purpose. (Romans 8:28)

We are Held

Natalie Grant's song *Held* (2005), sets to music a walk down the path of unreasonable suffering when something sacred is torn away, yet you survive. At every turn, what seems impossible to redeem becomes a blessing because we are "held" by an unseen hand. As the writer of Hebrews said concerning God's two-fold promise to Abraham,

> And so, after he had patiently endured, he obtained the promise. For men verily swear by the greater: and an oath for

> confirmation is to them and end of all strife. Wherein God, willing more abundantly to shew unto the heirs of promise the immutability of his counsel, confirmed it by an oath: that by two immutable things, in which it was impossible for God to lie, we might have a strong consolation, who have fled for refuge to lay hold upon the hope set before us: which hope we have as an anchor of the soul, both sure and stedfast, and which entereth into that within the veil; whither the forerunner is for us entered, even Jesus, made an high priest for ever after the order of Melchisedec. (Hebrews 6:15-20)

After Abraham obeyed the call to willingly offer his son, God promised to bless him and to multiply him that he might be a blessing to all the nations of the world (Genesis 22:15-18). In blessing God swore a promise, and He swore by Himself.

The Hebrew word *shaba* (meaning to swear, or seven, or complete) confirmed there exists nothing greater for God to promise by than by Himself. Therefore, the anchor of the soul holds within the veil both sure and steadfast because Jesus entered that veil.

The Siege of Bastogne

Battle of the Bulge

By early December 1944 during World War II, allied armies achieved great progress advancing toward remaining German strongholds. British and US troops entered the Ardennes Forest and the town of Bastogne in Belgium, though their supply line by now had been stretched thin. The circumstances developed into a classic case of over-confidence that fed into Germany's need for a strategic win.

In order to create a bargaining chip for peace concessions from Allied and Russian forces, Adolf Hitler took a bold gamble resulting in the Battle of the Bulge. To divide allied forces, Hitler consolidated an offensive force totaling over 410,000 troops, 1,400 tanks, 2,600 units of artillery, and 1,000 aircraft. His objective would be a thrust to reach the port city of Antwerp on the North Sea and put the Allies on the defensive. Initially called Operation Watch on the Rhine, the German offensive began on December 16 with the benefit of several strategic advantages.

In addition to thinly distributed Allied forces around Bastogne, a front of bitterly cold, snowy weather set in over most of Europe. Snow complicated the logistical support of Allied troops and grounded American aircraft. As German forces and equipment counter-attacked deep into the Ardennes, British and US troops could not be relieved by fresh soldiers. The situation appeared dire for the US 101st Airborne Division commanded by Brigadier General Anthony McAuliffe.

General George S. Patton Jr.

In order to relieve the siege of Bastogne, General Dwight Eisenhower met with his staff in Verdun on December 19, including General Omar

Bradley, Lieutenant General George Patton, British Field Marshal Bernard Montgomery, among others. General Eisenhower needed a bold, coordinated move to blunt the German offensive.

The only decisive proposal came from General George Patton. Patton, known for his off-the-cuff bluster, suggested allowing the Germans to reach Paris before cutting them off. Instead, Eisenhower directed Patton to take the lead and engage his army from the south in conjunction with the support of General Bradley.

Asked when he could initiate a move, General Patton replied, "As soon as you're through with me, I can attack the day after tomorrow morning." Dismissed as just bravado by most of his counterparts, Patton's plan had already been laid out with his own staff before the meeting.

The next forty-eight hours General Patton achieved what many describe as next to impossible. He turned his Third Army ninety degrees to the north and moved 100 miles in ice and snow, attacking December 22 as promised.

NUTS!

In Bastogne, German commander General Heinrich von Luettwitz sent a hand delivered letter to the US Army leadership post demanding immediate surrender. The response from General Anthony McAuliffe stated simply, "NUTS!". Von Luettwitz's threat of attack within two hours never materialized because German artillery had already been moved out of the area.

On December 22, Patton's army of 100,000 troops and 133,000 vehicles broke the German offensive and relieved Bastogne. This maneuver succeeded only because General Patton already made a plan and demonstrated the boldness to carry it out immediately.

Despite his often brash and abrasive nature, Patton demonstrated the leadership to get the job done. Don't give reasons; give results. Failure of the German counter-offensive called Operation Watch on the Rhine set the stage for unconditional surrender of Germany May 8, 1945.

Oskar Schindler

From 1944 until the end of World War II in 1945, an Austro-Hungarian industrialist named Oskar Schindler (1908-1974) began a clandestine campaign to save Jews from certain death. Schindler used his enamelware and ammunition businesses to employ Jews being sentenced to the death camps. By bribing German officials to overlook his efforts, he was able to spare about 1200 Jews near the end of World War II, at great expense to his personal safety.

Near the end of the war Oskar Schindler had expended his entire fortune equivalent to over a million dollars. In 1993 the Nation of Israel declared Schindler and his wife Emilie as "Righteous Among the Nations" in honor of their work shielding holocaust survivors. The honor recognizes non-Jews for their effort to protect Jews and the Nation of Israel. Director Steven Spielberg related the story of Oskar Schindler in the movie *Schindler's List* (1993).

Current events now foreshadow the soon coming of Jesus Christ and fulfillment of the Kingdom of God. Understanding the significance of these events is critical to discerning the times and seasons communicated in Scripture. The following chapters relate stories to frame the meaning of those events as they relate the blessed hope of Christ's return.

Shakespearean Play with Actors

Chapter Ten - All the World's a Stage

Over thirty years ago a mother strapped her two boys in the family van and drove to a lake. After exiting the vehicle, she allowed it to roll into the lake where it sank and drowned both boys. When the mother appeared on the news claiming her van had been carjacked and the boys taken, she appealed tearfully for their safe return.

As soon as the story broadcast on the news program, Debby declared, "She's lying." Frankly, I had the same suspicion because of her demeanor but did not want to judge before hearing the circumstances.

Upon investigation the police identified inconsistencies in the mother's statements. Specifically, the interrogating officer said she claimed someone carjacked her while stopped at a certain red light with no other cars at the intersection. As it turned out that light turned red only if a car approached the intersection from the cross-street.

When confronted with that fact, the mother changed her story claiming it must have been a different red light. At each telling her story changed when confronted with certain facts; a giveaway that the person is lying. Eventually she confessed to drowning her two boys because she became romantically involved with a co-worker who did not want children.

How did the investigators distinguish between a lie and the truth? By the logical principle of non-contradiction: It is impossible that two contradictory statements can both be true in the same sense. A changing story indicates duplicity. Often investigators plant a false narrative to see if the subject will change the story to adapt to their ploy. To maintain a fabrication requires remembering every detail of past lies.

In this case Debby's intuition detected deception because the mother's emotions and voice did not mirror someone whose children had just been kidnapped.

Theater Comedy and Tragedy Masks

Drama of Comedy and Tragedy

Beginning with Adam and Eve, everyone acts out drama on a personal or worldwide stage. Twentieth century theater and musical nihilists attempted to create plays or performances to challenge traditional structures using absurd themes that highlighted chaos rather than order. But even within chaos theory, patterns like fractals emerge demonstrating the impossibility of complete disorder.

As You Like It

In Shakespeare's play *As You Like It,* the character Jaques reflects on the emptiness of human achievement declaring, "All the world's a stage, and all the men and women merely players; they have their exits and their entrances, and one man in his time plays many parts, his acts being seven ages." Jaques then proceeds to list seven stages of life and their fleeting insignificance. Finally, he concludes, in second childhood a man is left without teeth, without eyes, without anything.

Shakespeare's words echo the closing argument of King Solomon when looking at the emptiness of events from a purely humanistic view.

> Vanity of vanities, saith the preacher; all is vanity. And moreover, because the preacher was wise, he still taught the people knowledge; yea, he gave good heed, and sought out, and set in order many proverbs. The preacher sought to find out acceptable words: and that which was written was upright, even words of truth." (Ecclesiastes 12:8-10)

In the purpose of God, everything reflects meaning. He fits everything into His plan, even what is sown in chaos.

> Let us hear the conclusion of the whole matter: Fear God, and keep his commandments: for this is the whole duty of man." (Ecclesiastes 12:13)

Church Scout Group Shooting a Revolutionary War Cannon

Chaos on the World Stage

Chapter 1 opened with a group of our church scouts at a mountain campfire scene. While telling the story of Jedediah Strong Smith, a grizzly bear interrupted the event only to be shot by the frontiersmen guarding the camp. Fortunately for the leader wearing the bear costume, it was staged to add realism and excitement reflecting the life of Jedediah Smith.

When beginning to work with the scout group, I asked one of the experienced leaders about the keys to successful events. He replied, "Focus three things: learning, fun, and achievement." Faced with unexpected chaos, Jesus spoke to His disciples as he calmed the sea, "Why are ye fearful, O ye of little faith." (Matthew 8:23-27)

Scouts Visiting a Jet Engine Training Center

Trebuchet Pumpkin Toss

Each of our church scout campout events for the Cincinnati Section included one or more projects to demonstrate skills and foster competition. The fall 2005 outing had a major challenge of doing a pumpkin toss using a catapult or trebuchet device. Scout groups had to design and build a throwing machine capable of launching an eight-inch diameter pumpkin the farthest distance.

Scripture records that King Azariah (806-739 BCE) [also called Uzziah], king of Judah had engines invented by cunning men to shoot arrows and great stones (2 Chronicles 26:15). These machines demonstrated the ability to breech city walls from a distance and were much more effective than simple battering rams at close range.

First, a short history of the siege engine. For thousands of years, armies have employed machines constructed to launch projectiles at the enemy or city walls. Some of these devices followed the principle of the bow or sling but at a much larger scale. Leonardo Da Vinci

(1452-1519) sketched a giant crossbow of over 80-foot width designed to shoot stones or beams, but the actual implementation of the principle proved impractical.

In the medieval period (500-1500 CE), invention of the counter-weighted trebuchet greatly increased the ability to throw large projectiles over long distances. A trebuchet differs from a catapult in that it lengthens the effective "throwing arm" by the addition of a rope and pouch for the projectile. The same principle enables a slingshot to extend the thrower's arm and thus increase the projectile velocity.

King Edward-1 and his War Wolf Trebuchet

King Edward-1 of England (1272-1307) commissioned the largest trebuchet ever built with a 20-ton counterweight and a 50-foot throwing arm. Named the War Wolf, this machine could launch 300-pound stones a distance of 900 feet to demolish castle walls. Occasionally a trebuchet would be employed for psychological warfare by launching flaming projectiles, diseased animals, or severed heads of captured enemies over city walls.

Articulating Pivot Trebuchet Throwing a Pumpkin

Our scout group chose to create a specialized trebuchet with an articulating pivot. This means the pivot point of the twelve-foot throwing arm would rotate forward similar to a pitcher's motion releasing a baseball. In order to prove out the principle, a one-fifth-scale model was designed to demonstrate the device would operate as planned. The model performed perfectly and launched the ball bearing projectile over sixty feet.

Because the throwing distance does not scale by the length of the arm, the full-scale trebuchet distance could not be calculated easily. But the objective was to achieve a throw of over two hundred feet.

One of the requirements of this competition called for use of natural materials rather than commercial hardware. That meant finding a few straight, live trees suitable for the throwing arm and supports. Green ash wood has very high strength and flexibility with few knots. Several long, straight ash trees were located to create the trebuchet structural supports and twelve foot arm (3.7 meter). Assembly of the support frame used rope lashing rather than mechanical fasteners.

A major consideration to gain maximum throwing arm velocity

involved making the moving parts light, yet as strong as possible. The lighter the throwing arm, the less energy required to move it, meaning more energy available to accelerate the pumpkin. The day of the competition our trebuchet launched a pumpkin 218 feet for the winning throw.

Derby Krash-o-Rama

Each year our church scouting program conducted a derby car race competition. For this event the girls also participated. The boys and girls received a kit including a block of wood, wheels, and axles with a challenge to create a car based on a specific racing concept. A derby race theme might be NASCAR, police, movie cars, Formula 1, antique, or military vehicles.

Derby Model of Paladin 155 mm Howitzer Firing 22 Caliber Blank

Cars could be created for racing, for show, or both with trophies given according to the age group. The competition encouraged boys and girls to learn skills necessary to complete a project with the objective of winning a racing contest.

Scout leaders and parents assisted by rough shaping a car body, but the boys were expected to choose the car style and do the finish work. Race day involved a frenzy of activity with setup of race tracks, final registration, and judging of the show cars. After the first round of racing about half of the boys and girls had been eliminated. The rest continued on to race in finals to select the trophy winners. Unfortunately, many of the disqualified racers and parents would leave without waiting for final runs and awarding of trophies.

Kandi's Krash-o-Rama Event Crash Dummies

In order to address the issue, our Cincinnati Sectional Commander Gary Bayes asked a few of the leaders to come up with an alternative competition. The object would be to create a race event enabling those who lost to compete for a consolation trophy. They would therefore be more likely to remain until the award ceremony. After some brainstorming we developed a race called Kandi's Krash-o-Rama.

Racers who did not get to the finals had the opportunity to send their car down a racetrack and hit a weighted crash dummy sled. The one driving the crash sled the farthest down the track won a trophy. First round qualifiers were not eligible for Krash-o-Rama. Not surprising, the event became a big success because most of the racers and parents stayed for the full event in order to participate in Kandi's Krash-o-Rama.

As an allegory to the story about ten maidens and their lamps in Chapter 1, it is important to be present and prepared for the awards.

Roman Soldiers Killing Galileans in the Temple

Chapter Eleven - Pathos, Ethos, Logos

Breaking News Alert! Pilate Executes Galileans Mid-Sacrifice! Being first on-scene with the latest news is thrilling. It helps to be able to frame what the event means. Unfortunately, sometimes the message misses the point. Jesus responded to a report of an attack by Pilate's soldiers by correcting His follower's assumptions (Luke 13:1-5).

If they presumed death of the Galileans meant they somehow deserved to perish, Jesus said,

> I tell you, Nay: but, except you repent, ye shall all likewise perish. (Luke 13:3)

Anecdotes like this in Scripture often lack context. Galilee gained a reputation as a hotbed of political unrest. As home to Jesus as well as the zealots, the land of Galilee fomented ongoing disturbances. Galilean zealots rioted against the corrupt Jewish leadership, governor Pontius Pilate, and Roman rule. Scripture identifies one apostle, Simon Zelotes (Luke 6:15), likely sympathetic to the cause of the zealots.

Under pressure from Rome, Pilate eventually acted to quell the source of opposition. In this case, rather than sending a band of soldiers north into Galilee, he did so by transmitting a lethal message to a group of Galilean zealots who gathered in the temple courtyard for traditional sacrifices.

The Greeks

In addition to zealots, Hellenists throughout the Roman empire represented another prominent segment of the Jewish population. These Jews observed the law yet embraced Greek and Roman culture. Often referred to as Greeks, the Hellenists used the Greek Septuagint translation of the Old Testament. Stephen and Philip were two well-known Hellenist followers of Jesus.

After the triumphal entry into Jerusalem on Palm Sunday, a group of "Greeks" approached Philip asking to see Jesus (John 12:20-32). These Greek Jews came to worship in Jerusalem at the Passover. When Jesus heard from Philip and Andrew that a group of Greeks sought to see Him, it appears Jesus addressed that as a sign.

> And Jesus answered them, saying, The hour is come, that the Son of man should be glorified. Verily, verily, I say unto you, Except a corn of wheat fall into the ground and die, it abideth alone: but if it die, it bringeth forth much fruit. (John 12:23,24)

Jesus used the approach of Greek Jews to announce that the hour of His sacrifice had come, as He declared,

> Father, glorify thy name. Then came there a voice from heaven, saying, I have both glorified it, and will glorify it again. (John 12:28)

Scripture contains passionate anecdotes like this scattered throughout both New and Old Testament narrative. Not only does Scripture include a moral framework and rational development but it is punctuated with emotional stories. In Sunday School class someone once suggested the Bible would make more sense formatted as an instruction manual; logical but not so provocative. Thankfully the Bible does not read like a maintenance manual or it would be read less than it is already.

Greek philosopher Aristotle (384-322 BCE) listed three primary threads of persuasion (in Greek); *pathos, ethos*, and *logos*. These apply to speaking, writing, art, theater, and any form of communication. Pathos or passion appeals to the emotions while ethos addresses moral character and logos draws on the intellect by reasoned truth. Lacking any of the three elements, the message falls flat.

Spiritual Apathy

An early teaching of the church regarding moral character developed around the principle of seven deadly sins. Apart from the list of ten commandments, early church fathers sought also to identify the most grievous misbehaviors.

Tertullian (about 160-220 CE) created a list of seven sins based largely on the ten commandments (idolatry, blasphemy, murder, adultery, fornication, false witness, apostasy). During the sixth century, Pope Gregory I articulated seven well-known virtues listed following along with their original Latin names.

1. Humility (*Humilitas*)
2. Charity (*Caritas*)
3. Chastity (*Castitas*)
4. Gratitude (*Humanitas*, or *Gratia*)
5. Temperance (*Temperantia*)
6. Patience (*Patientia*)
7. Diligence (*Industria*, or *Diligentia*)

By contrast Pope Gregory I also identified seven deadly sins, recorded with Latin names. The list provided a Latin medieval acronym *SALIGIA* as a reminder of the sins.

1. Pride (*Superbia*)
2. Greed (*Avaritia*)
3. Lust (*Luxuria*)
4. Envy (*Invidia*)
5. Gluttony (*Gula*)
6. Wrath (*Ira*)
7. Sloth (*Acedia*)

The focus here concentrates on the seventh, commonly labeled "sloth". Rather than referring to laziness, the Latin word *acedia*, would be best translated as "spiritual apathy". Tertullian characterized this as apostacy. More than any other failing of the human condition, spiritual indifference lies at the root of alienation from God.

Spiritual apathy traces back to the garden where God blessed Adam and Eve and gave them a directive.

> And God blessed them, and God said unto them, Be fruitful, and multiply, and replenish the earth, and subdue it: and have dominion over the fish of the sea, and over the fowl of the air, and over every living thing that moveth upon the earth." (Genesis 1:28)

In addition, God placed Adam in the garden of Eden with the command "to dress it and keep it." (Genesis 2:15) Finally, God brought every creature to Adam that he might name them, and "whatsoever Adam called every living creature, that was the name thereof." (Genesis 2:19) Therefore, Adam received a three-fold work assignment and he appeared to have carried out the last two at the expense of the first which was to be fruitful and multiply. Eve on the other hand found herself with a lot of free time.

Good and Evil

The only warning with consequences to Adam pertained to not eating the fruit of the Tree of the Knowledge of Good and Evil. Death as punishment must have had meaning to Adam because he stayed away, at least until Eve fell for the serpent's lie;

> And the serpent said unto the woman, Ye shall not surely die; For God doth know that in the day ye eat thereof, then your eyes shall be opened, and ye shall be as gods, knowing good and evil. (Genesis 3:4,5)

God reprimanded each and specified consequences to Adam and Eve which passed down to their descendants. With the birth of Cain and Abel, both evidently understood they must make an offering to God. Abel, a herdsman, offered a blood sacrifice of the firstling of his flock. By contrast Cain (Hebrew meaning to possess), a tiller of the ground, brought the fruit of his labors.

Abel and Cain Offer Their Sacrifices

The Hebrew word for tiller (*Obed*) means worker or servant. God's challenge to Cain implied he should have known the requirement and done accordingly by obtaining a lamb from Abel. Disobedience and apostacy continued until God directed Noah to build an ark to save himself and his family (Genesis 6).

After the flood the lineage of Ham produced Cush and his son Nimrod, builder of the tower of Babel (Genesis 11:9). Nimrod also founded the city of Babylon in the plain of Shinar (Genesis 10:9-11). From Genesis to Revelation, Babylon (Hebrew meaning confusion) represents the chaos from substituting other things for relationship with God. In the last chapter, the themes of this book climax in Act III and the final scenes of a play titled, *L'Acedia – Land of Leisure.*

Scene 1 opens to chaos after a period of unrest and rioting in L'Acedia. The drama in the following scenes unfolds to paint a picture of events described in Revelation 6 with the opening of seals on a roll (book).

In each of the scenes, characters seek to control their own and other's destiny. But powerful individuals, organizations, and invisible forces manipulate to their own ends. Behind the stage a Jester pulls the levers for his purpose. And unseen to all, Providence works everything to His ultimate plan.

Jo'an d'Acosta in the Court of Tsar Peter-I

Chapter Twelve - The Jester

FEW ENTERTAINERS SUCCESSFULLY exploit the art of comedic insult. Schlemiel and schlimazel (Yiddish for bumbler and victim) served as comic relief among European diaspora Jews (Ashkenazim). Faced with tragedy, this coping strategy enabled them to find humor in the midst of misfortune.

A classic model of the practice would be comedian Don Rickles (1926-2017), affectionately labelled "Mr. Warmth" for his unrelenting barrage of satirical "compliments". Born to a Jewish family in Queens, New York, Rickles perfected his rapid-fire style of insult humor as a standup comedian by confronting spectators in the audience. Even his detractors enthusiastically embraced Don's quick-witted retaliation against hecklers.

As I explained in the final chapter of the book *Rend the Heavens ...and Come Down*, mocking sarcasm grows out of the medieval court practice of appointing a court jester. In the pompous atmosphere surrounding royalty, the court clown created a great contradiction.

Though playing the part of fool, he was traditionally more articulate than anyone else in attendance. At times performing as a minstrel, the Jester would set satire to music. So, he played his hand skillfully to keep everyone entertained, distracted, and off-balance.

Jester Jo'an d'Acosta

One of the most successful jesters, Jo'an d'Acosta (1665-1750) served in the royal court of Tsar Peter-I of Russia. Though born in Holland, d'Acosta spoke multiple European languages. After a successful career working as a Jewish broker in Hamburg, he moved to Saint Petersburg, Russia where he received an appointment as royal jester to the Tsar in 1714. Tsar Peter-I enjoyed debating philosophy and theology with d'Acosta who had an excellent command of Scripture.

The role of jester can be traced back to ancient Egypt when Pharoah often employed a talented foreigner to disrupt the status quo. By the Middle Ages a court joker or minstrel evolved as a common fixture for royal entertainment. In the mid-nineteenth century use of a wild or trump card in gamesmanship mirrored the character of a court jester.

The Jester could always trump an opponent, offering an element of surprise to a normally predictable contest. Having no official position or power of his own, the court fool amused by out-maneuvering the powerful with his verbal sleight of hand. A joker's outlandish appearance assaulted imperial sensibilities.

When someone presumed to unravel his ruse, the clown would in turn find a way to catch the unwary victim off balance. As royal theater played out, one would expect the king to eventually rein him in and bring the court to an orderly conclusion. It seemed everyone played by the rules, but not the Joker. To both crown and court, he served a very important purpose; like Shakespeare's villain Iago in the play *Othello* or the "Motley Fool" Touchstone in *As You Like It*.

Shakespearean Theater

Iago, A Perfect Villian

Most of William Shakespeare's thirty-seven plays include an antagonist. In *Othello*, writer William Shakespeare (1564-1616) created an adversary Iago, featuring what has been termed the perfect villain. The name represents an abbreviation of the Spanish name Santiago (or English St. James). In turn, the root derives from Yaakov or Jacob (Hebrew meaning deceiver).

Though not a traditional jester, Iago fills the role of adversary with a subtle twist. Iago commits no crime, but with deception and enticement he incites others to satisfy his ill intentions. By subtle monologues to the audience, Iago plots and draws the listener into his conspiracy. In a final triumph over evil, the conclusion of the play *Othello* has the wife of Iago betray him resulting in his execution.

The agenda of Satan includes enticing as many as possible into his scheme, and persuading them to do the deeds.

Touchstone, The Wise Fool

Shakespeare's *As You Like It* contrasts the vanity and corruption of King Frederick's court with the simplicity of pastoral life in the Forest of Arden. When characters are exiled by the new king to Arden, they escape the court's hypocrisy only to face naïve duplicity in the forest.

Touchstone, labeled the Motley Fool for his patchwork of jester's clothing, follows them to Arden. There Touchstone trades his role addressing the vanity of royalty with the task of correcting simple-minded assumptions of rurality. Though he cynically dismisses marriage, Touchstone agrees to marry Audrey, a goat-herder for "practical" reasons. With a flourish of marriages and restored relationships, Shakespeare ends the play on a positive note.

Jacob Wrestling All Night with an Angel

The biblical story of Jacob describes how he tricked his brother Esau out of the birthright (Genesis 25:29-34) and his blessing (Genesis 27:18-29). Thus, in fulfillment of the prophesy at Jacob's birth, he was named Yakob (deceiver). God in turn rejected Esau for selling his birthright to Jacob for a bowl of bean soup.

Twenty years after separating, Jacob and Esau prepared to meet again. Fearing the worst prior to confronting Esau, Jacob wrestled all night with God (Genesis 32:24-32) who manifested as a man or an angel (Hosea 12:3,4). Upon prevailing with God, the angel renamed Jacob as Israel (meaning, as a prince he will rule with God).

The story of Jacob (or Israel) and Esau portrays God's purpose of working everything to His plan, even in the face of human failure and adversity. In the end, the only loser will be God's villain Heylel, and anyone aligned with him.

Negotiation Strategy

Effective Negotiation

In 1974 Robert Ringer published a book, *Winning Through Intimidation.* The title represents somewhat of a misnomer. Ringer speaks in his book about winning "against" rather than winning "by" intimidation. Instead of falling victim to intimidation in a business negotiation, Ringer advised how to posture in order to avoid being intimidated. Business discussions do not need to be seen as win-lose propositions.

After promotion to a managerial position at the company, the training department assigned me to a week-long series of courses for leadership skills. One of the classes was titled "Effective Negotiation". Leadership negotiation involves interaction with peers and subordinates as well as business transactions with external suppliers. Often the process takes the form of a clash of wills, but the reality is complex.

The negotiation class included several team activities built around competitive scenarios. One exercise had nation-state teams paired with opposing nation-states in a conflict discussion. Each team and their adversaries received assets and armaments along with rules of engagement and outcomes for cooperation or conflict. At the conclusion of each negotiation phase, members of both sides decided among themselves whether to collaborate or clash with the opposing nation. According to the rules, opponents won or lost assets and armament depending on whether their team chose to cooperate or attack.

Not surprising, most of the participants viewed the competition as "winner takes all", even if both sides lost almost everything. Instead, rules of the exercise favored collaboration over conflict. The only way to win required both sides to eventually vote for cooperation at each phase. If one or both chose to attack, they all lost even though attacking first resulted in losing less.

Out of fear of being attacked, most voted to attack first. It took several rounds of votes before most of the team members realized the only way to "win" required players to convince their own stubborn teammates to vote for cooperation. If a team attacked another, trust was lost. Negotiation training emphasized good faith discussions rather than winning against an opponent at all costs.

Colonial Game of Chess

The negotiation exercise demonstrated that agreements made in good faith can work for mutual benefit, much like a game of chess. When one or both parties seek to dominate and win, lost trust can be difficult to regain. Abimelech (Hebrew – my father is king, his name and title) sought to negotiate peace with Abraham (Genesis 21:22-32) because he felt threatened.

The resulting covenant of Beersheba (well of the oath, or seven, or completion) provides a foreshadowing of the soon coming seven-year covenant between Israel and the nations. Similarly, when Isaac increased in riches, he negotiated an agreement of protection with Abimelech (likely son of the first Abimelech). In Chapter 17 these agreements set the stage for the return of Jesus Christ.

The Angel Heylel Before His Fall

The Fall of Heylel (Lucifer)

God did not have a pre-existing opponent. Archangel Heylel (Hebrew meaning Morning Star or Latin name Lucifer) was created by God and placed in position as the leader of worship in heaven. In the case of God's Kingdom, His adversary Heylel envisioned himself prevailing and rising above God.

There is no negotiation with Heylel. In order to accomplish

his objective, Satan or Heylel fostered conflict against God, among believers, and against the church. In Heaven, one-third of the angels joined his rebellion against God and they were cast out. Now, his end game seeks to hold off inevitable destruction as long as possible.

> How art thou fallen from heaven, O Lucifer [Heylel], son of the morning! How art thou cut down to the ground, which didst weaken the nations! For thou hast said in thine heart, I will ascend into heaven, I will exalt my throne above the stars of God: I will sit also upon the mount of the congregation, in the sides of the north: I will ascend above the heights of the clouds; I will be like the most High. Yet thou shalt be brought down to hell, to the sides of the pit. They that see thee shall narrowly look upon thee, and consider thee, saying, Is this the man that made the earth to tremble, that did shake kingdoms; That made the world as a wilderness, and destroyed the cities thereof; that opened not the house of his prisoners? (Isaiah 14:12-17)

Observers of the fallen Heylel in this passage marvel at his lack of presence. It would seem a man who troubled the earth and shook kingdoms would have a formidable appearance. Instead, his form and demeanor emerge like your worst blind date.

The Conflicted Clown

Put on a Happy Face

Before his fall from Heaven, Heylel held an exalted position as leader of music and worship. His splendor and pride caused his downfall.

> Thou art the anointed cherub that covereth; and I have set thee so: thou wast upon the holy mountain of God; thou hast walked up and down in the midst of the stones of fire. Thou wast perfect in thy ways from the day that thou wast created, till iniquity was found in thee. By the multitude of thy merchandise they have filled the midst of thee with violence, and thou hast sinned: therefore I will cast thee as profane out of the mountain of God: and I will destroy thee, O covering cherub, from the midst of the stones of fire. Thine heart was lifted up because of thy beauty, thou hast corrupted thy wisdom by reason of thy brightness: I will cast thee to the ground, I will lay thee before kings, that they may behold thee. ...All they that know thee among the people shall be astonished at thee: thou shalt be a terror, and never shalt thou be any more. (Ezekiel 28:14-17, 19)

The Fortune Teller

Earlier in Chapter 5, I alluded to the story of Norma Zimmer, featured singer for several years on The Lawrence Welk show. My parents regularly viewed the television show where Norma sang as the "Champagne Lady". In 1972 she published her life story *Norma* relating the incident that nearly destroyed her family.

A seemingly innocent decision by her father to stop at a gypsy fortune teller became an event that turned a normally difficult upbringing during the Great Depression into a trail of traumatic experiences. Prior to her birth, Norma's parents took a Sunday stroll along a country road where a band of gypsies camped. There they were confronted by a lady who offered to tell her father's fortune.

Thinking it would be harmless entertainment, Norma's father Peter Larson allowed the woman to read his palm. Upon observing the work injury to Peter's hand, the gypsy palm reader expressed surprise, and to his amazement, accurately described a history of painful events in his life. The events she spoke may have been true, but her interpretation of the meaning of those events represented a lie.

Now his ears were opened. She then predicted that Norma's mother, who was pregnant, would give birth to a daughter which was Norma. Further, the gypsy predicted several dark forecasts for her father that would continue to make him unsuccessful.

With his attention now riveted, she concluded by laying a heavy millstone upon Peter Larson, saying his son Max would attempt to kill him. That dark cloud hung like a curse on Norma's father. His problem with alcohol blossomed into addiction that later extended also to Norma's mother.

Every personal setback or drinking bout ended with Norma's father wondering when his son would try to take his life. What started as seemingly harmless curiosity developed into a cancerous growth that consumed Norma's family for decades. It was only later in life that her father Peter Larson found forgiveness.

In her book, Norma alluded to their difficult family situation while living in the Pacific Northwest. During the 1930s depression she described the threat of a communist political takeover in the area. As a teenager raised in the vicinity of Seattle-Tacoma, Washington, Norma heard the gospel message and committed her life to Jesus Christ. Later, during her career singing on The Lawrence Welk show, she received an invitation to sing during a Billy Graham crusade, and continued to perform in that capacity for many years.

In the 1960s, while my extended family gathered Saturday evening to watch The Lawrence Welk Show, one of my close relatives would often do fortune-telling. At times, friends, family, and sometimes strangers would arrive and sit for a tea leaf reading in the kitchen. Occasionally I would overhear a surprised response about something spoken, or there was an exclamation about an event that came true. There would then be an exchange of money for the reading.

Looking back, what at the time may have sounded true to the hearer, instead became a curse. As so often occurs, trouble does not arise from outside but as a result of poor judgment; the tyranny of unintended consequences. As noted in Chapter 7, trouble or tribulation represents a door of hope (Hosea 2:14,15).

What's In a Name?

A biblical principle surrounds the use of names. In the Hebrew language and Jewish culture, use of a name describes behavior seen or qualities desired. The same notion of names applies to Arabic words and Muslim society. Hebrew and Arabic are both languages of semitic origin. Therefore, Biblical names generally carry meaning and a message. That is why it is informative to look at the original Hebrew definition of names to understand the significance as it relates to how that person fits into the biblical story.

While working at the company as a representative for the LM6000 industrial gas turbine, I had opportunities to travel the world to assure proper installation and startup of the engine we developed in the 1990s. Several of those power plants in India required trips to resolve issues after initial startup and commissioning. In one case, a pair of LM6000s had been installed for a dedicated power plant at an Indian petrochemical company.

The engines had been fitted with fuel nozzles designed for operation on naphtha residual liquid, a waste by-product of petrochemical processing. Several trips were required to identify and resolve a cavitation erosion problem with the special fuel nozzles.

During a visit in the Indian State of Gujarat, I managed to take photographs of the LM6000 turbine engines, power plant, and construction work on the facility. Due to the lack of concrete equipment, most of the mixing and pouring was completed by hand using a crew of women carrying pans of concrete to the location for finishing. While taking photographs, many of the women smiled for the camera. Later that day the construction engineer on-site cautioned me that these women were Khoja Shia Muslims, and by posing for the camera they would possibly be beaten.

Shia Muslim Women in India Pouring Concrete

The State of Gujarat in India includes a large number of Twelver Shia Muslims of the Khoja persuasion. In addition to common Shia Muslim teachings, Twelvers believe in a succession of twelve Imams descended from Muhammad, eleven of whom have already appeared. The first Imam was Ali ibn Abi Talib and the final yet to come, Muhammad al-Mahdi believed to be in "occultation" or hidden from view until his reappearing. According to Khoja Twelver theology, the revelation of the twelfth Imam will be a return of the last Imam Muhammad ibn Hasan who was born 868 CE and did not die but entered a state of concealment in 941 CE.

At his return as Imam Mahdi, Khoja Muslims believe he will come to impose worldwide submission to Islam and usher in world peace. Also, they teach the Madhi will be accompanied by the second coming of Jesus Christ (or Isa in Arabic) whom they acknowledge as a prophet rather than the Son of God.

In their teaching, Jesus Christ will return to confront the antichrist figure al-Masih ad-Dajjal (Arabic meaning Messiah-False). All these characteristics of Islam mirror the Biblical story of God's final chapter, but with a reversal of roles to fit their narrative.

The recent rise to power of New York City mayor Zohran Kwame Mamdani, a virtual unknown, presents an interesting premonition based on the meaning of names. An immigrant from Uganda, Zohran Mamdani descends from a clan of the Indian Twelver Khoja Muslims. In Arabic, *Zohran* means morning star or to shine (equivalent to the Latin name *Lucifer* and Hebrew name *Heylel*). The meaning of *Kwame* is born on Saturday, or seventh day.

Mamdani (Arabic meaning to be praised) derives from the name Muhammad, though he makes no claim to be a direct descendant. How this person figures into coming events at the return of Jesus Christ remains to be seen. Zohran Mamdani's stance is openly anti-Israel and antisemitic.

Though labeling himself a democratic socialist, Mamdani's far-left policies have been characterized as outright communist. Given Mamdani's refusal to condemn the radical phrase "globalize the intifada", it is disturbing to see that fully one third of New York City Jews helped to tip the scale and vote him into office. The parable of Jotham portrays the toxic outcome resulting from a dysfunctional alliance.

Jotham's Parable of the Trees

The biblical story of Gideon (Judges 6) relates his transformation from a timid young man to a great warrior. Called of God to lead the men of Israel against a much larger army of Midianites, Gideon placed two "wool fleece" tests before God and thereby mustered the courage to attack the army of Midian.

When God tested Gideon's Army, the Lord reduced his men from thirty-two thousand down to ten thousand. As a final test, they drank at the brook and only 300 footmen who lapped with their hand were chosen to confront the superior force of Midian.

After defeating the Midianites with 300 men, Gideon ruled Israel as judge for forty years. Upon his death, Gideon left seventy sons from whom they would need to choose a successor. In addition to sons of his multiple wives, Gideon also had a son Abimelech of a concubine from the Canaanite city of Shechem.

Before a replacement ruler or judge could be chosen for the whole nation, Abimelech rallied support to himself among the men of Shechem where he had been raised. Once Abimelech received backing from the City of Shechem, they allocated finances so he could hire a band of vain followers.

Abimelech's first order of business concerned the elimination of competition by his seventy half-brothers. With the assistance of the men of Shechem, all his brothers were killed with the exception of Jotham the youngest son. The Hebrew meaning of Jotham is YHWH is perfection (or Yahweh is completion).

Upon his escape, Jotham climbed Mount Gerizim (hill of blessing) opposite Mount Ebal (hill of cursing), both of which overlook the City of Shechem. Jotham proceeded to tell a parable regarding the apparent contradiction of their choice of leadership.

The Parable of Jotham on Mount Gerizim

Despite the courage and sacrifice of Gideon, the men of Shechem selected Abimelech, an outcast foreigner, as the leader and judge rather than one of Gideon's legitimate Jewish sons. A very important observation: as son of the Canaanite inhabitants of Shechem, Abimelech would have been from the worshippers of the false god Baal-Berith (Hebrew for lord of the covenant). When attacked, the men of Shechem fled into the house of Baal-Berith (Judges 9:46).

Notice also, Abimelech was the name of the Canaanite King(s) with whom Abraham and Isaac made a covenant of protection. In the case of Abraham, he set apart seven ewe lambs as a token of the covenant at Beersheba (Hebrew for well of the oath, or seven, or completion). The current Abraham Accords seek to establish a similar covenant for peace between Israel and her adversaries.

Once chosen, Abimelech and the men of Shechem elected to slay all of his half-brothers, though Jotham escaped to tell the story. In his parable, Jotham spoke of the trees going out to choose a king over them. In turn the trees offered the position to the olive tree, the fig tree, the grape vine, and finally a briar bush. Each of the first three were preoccupied with their fruit or function. Having no time to govern, they dismissed the offer. But the unfruitful briar bush agreed to rule provided the trees put their complete trust under the briar's shadow.

Jotham closed his parable by pronouncing a conditional curse on Abimelech and the men of Shechem. He declared, if Abimelech and the men of Shechem had dealt truly and sincerely with the house of his father Gideon, then may they rejoice in each other. But if they had dealt deceitfully, Jotham pronounced that fire would come out from Abimelech and devour the men of Shechem and vice versa. Indeed, the curse of Jotham took effect after three years when God sent an evil spirit to create division among them, resulting in mutual destruction.

In the battle, Abimelech destroyed the men of the City of Shechem. Next Abimelech set his sights on the neighboring city of Thebez. While assaulting the wall of Thebez, a woman on the wall threw down a piece of millstone and crushed the skull of Abimelech. Not wanting to be slain by a woman, Abimelech asked his armor bearer to slay him with his sword.

Abimelech Killed by A Piece of Millstone

What is the moral of the story? According to God's perfect plan (Hebrew Jotham – YHWH is perfect), His purpose always prevails while evil destroys itself. How ironic that twenty-five years after the 911 Muslim terrorist attack on New York City, Jewish leaders openly embraced a mayoral candidate who refused to condemn global intifada against Israel and the Jews. Greater Metropolitan New York City includes about two million Jews, the world's second largest municipal Jewish population next to Tel Aviv, Israel.

In the case of Abimelech and Shechem, after three years of relative calm, the curse took effect. So also, when Elijah imposed a drought (1 Kings 17:1), he called for rain after the third year (1 Kings 18:1). Jesus declared (Luke 4:25) and James concurred (James 5:17) that in the days of Elijah, the heavens were shut up for three years and six months before Elijah prayed for rain.

According to the prophet Daniel (Daniel 9:27), in the middle of a seven-year period (a Hebrew *shabua*), a prince to come (or Antichrist) will break a covenant agreement. That historical event, both Daniel and Jesus described as the Abomination of Desolation. Daniel's prophesy was fulfilled when the temple was defiled under the reign of Greek king Antiochus-IV Epiphanes in December 167 BCE. Jesus spoke in the year 33 CE of a similar event yet to come.

Immediately following a future Abomination of Desolation, an Antichrist figure will break his covenant agreement and engage the world with a period of great tribulation. The Jewish people, Israel, and the church of believers will be subject to Satan's effort to destroy the hidden one-hundred-forty-four thousand Jews in the hope of holding off his own eventual demise.

On a particular occurrence of Rosh Hashanah (Festival of Trumpets), Jesus Christ will open the sixth seal and initiate His return for the believers. This transpires about eighteen months after the Abomination of Desolation. No one knows the hour or day, though we do know the corresponding events with the times and seasons. Be prepared!

Antiochus-IV Epiphanes' Statue of Zeus in the Jewish Temple

That seven-year period of 2,520 days, according to the 360-day Jewish calendar, corresponds to the exact number of days between the formation of an agreement for the protection of Israel and the final day of that period, Yom Kippur. On that day Jesus Christ will return bodily to one-hundred-forty-four thousand Jews hidden in the Jordanian wilderness of Petra (or Sela). In the fulfillment of the Day of Atonement they acknowledge Jesus as Messiah.

During the rule of Antiochus-IV Epiphanes, the king made an agreement with Hellenistic Jews for their protection provided they abandon Jewish customs. Antiochus-IV later violated the agreement by placing a statue of the Greek god Zeus in the temple with his own image and sacrificing a pig on the altar.

The priest Mattathias then led an uprising and three years later under the leadership of his son Judah Macabee, the Greeks were overthrown. Cleansing and rededication of the temple on 25 Kislev in 164 BCE instituted the Festival of Lights (Hanukkah). This prefigures a future rest beginning the millennial reign of Christ.

All of these historical events of the Maccabean revolt are recorded by Josephus in the *Antiquities of the Jews* as well as the two books of *I and II Maccabees* from the Apocrypha. The Abomination of Desolation during that period and resulting tribulation only foreshadow the corresponding events when the Antichrist initiates the time of great tribulation.

In Chapter 18 of this book, Act III of the play *L'Acedia Land of Leisure* follows the final desperate attempt of the Jester to manipulate the script and steal the crown.

Jester Carrying off a 9-Stem Hanukkah Candle

City of Babylon Wall Illustration

Chapter Thirteen - Last Train Out of Babel

THE SEVENTEENTH CENTURY agricultural region of Huntington South on Long Island, New York traces its history to 1689. In 1710 Captain Jacob Conklin built the first home in what later came to be called the Town of Babylon, New York. Captain Conklin's son Colonel Platt Conklin served in the Revolutionary War and later settled in the area after marrying Phebe Smith.

Two decades after the end of the Revolutionary War, their son Nathaniel Conklin returned to the area with his family and mother Phebe. In 1803 Nathaniel constructed a home in what was then named Huntington Town.

The Old Train Station of Babylon, Long Island, New York

Upon learning her home would be next to a tavern, Phebe Conklin christened the place "Babylon" after the fallen Biblical City of Babylon. In order to change the narrative, Nathaniel Conklin had a cornerstone for their home engraved with the name "New Babylon". The citizens of the community eventually adopted the name for the Town of Babylon in Long Island, New York.

The ancient City of Babylon began as a variant of the name Babel (meaning confusion), conferred upon the city and tower built within the plain of Shinar (Genesis 11:9). After the flood God's mandate called for mankind to go forth and populate the earth. From Noah's three sons, Shem, Ham, and Japheth were born sons after the flood reflecting three tribes.

> And God blessed Noah and his sons, and said unto them, Be fruitful, and multiply, and replenish the earth. (Genesis 9:1)

The Biblical Tower of Babel

Instead, the people conspired to create a tower that reached to heaven. As a consequence of their self-will, God confused the tongues and each language group scattered themselves throughout the earth. Of Noah's son Ham were born four sons; Cush, Mizraim, Phut, and Canaan. Cush in turn had six sons; Seba, Havilah, Sabtah, Raamah, Sabtecha, and Nimrod.

It was Nimrod who founded Babel and other great cities in the Mesopotamian plain called the Land of Shinar. The description of Nimrod paints him as a mighty hunter (Hebrew *gibbor sayid*) against YHWH (or Yahweh).

Nimrod, Founder of the City Babel

Use of the expression "mighty hunter" does not reflect a valiant warrior but rather a proud tyrant contrary to YHWH. The depiction of Nimrod describes a builder of a kingdom and adversary to the plan and purpose of God. Throughout the biblical narrative, the ancient City of Babel, the Kingdom of Babylon, and mystical Babylon serve as a contrast to God's purpose in Jerusalem and the Kingdom of God. Like Nimrod, throughout history mankind and nations reflect a struggle against God's plan.

Evil Serves God's Purpose

Why does God continue to allow evil as a contradiction to his plan? When Jerusalem fell to King Nebuchadnezzar of Babylon in 586 BCE, he exiled most Jewish leaders to the City of Babylon (Jeremiah 39). Over the final two-year siege, Jeremiah prophesied to King Jehoiakim, King Zedekiah and the Jewish rulers of Judah that they should submit to the Babylonians. At the reading of Jeremiah's prophesies, Jehoiakim had the roll cut and burned in the fire. The leaders viewed Jeremiah as a traitor based on their belief they would be delivered from the Babylonians.

After all, Israel was delivered from Sennacherib 175 years earlier (Isaiah 37:21-38) when the prophet Isaiah assured the leaders of Jerusalem He would destroy the Assyrians. In Jeremiah's time King Jehoiakim and King Zedekiah deceived themselves, assured that God would intervene as He did at the prophesy of Isaiah to King Hezekiah. This deception characterizes the warning of Jesus regarding the days before His return (Matthew 24:9-13).

Upon Jerusalem's fall, those who conspired to resist the Babylonian takeover were executed by Nebuchadnezzar in Riblah of Syria (2 Kings 25:18-21). Only the poor populace and a limited staff of leaders remained in Jerusalem after the city and temple had been burned. Nebuchadnezzar's army chief Nebuzaradan received a directive to allow the prophet Jeremiah to remain in Jerusalem or go to Babylon. Jeremiah chose to stay with the remnant rather than exile to Babylon,

Jehoiakim Has the Roll of Jeremiah Burned

The last of three groups then exiled from Israel to Babylon. There they remained seventy years. When the Medo-Persian empire overthrew Babylon, Cyrus King of Persia ordered the funding for Jewish exiles who volunteered to return in order to rebuild Jerusalem.

During the seventy-year exile prophesied by Jeremiah (Jeremiah 25:11-12 and 29:10), those in Babylon prospered including Mordecai, Daniel, and his fellows who received positions of leadership under Nebuchadnezzar and Ahasuerus. Daniel continued his role under King Darius and King Cyrus after the Medo-Persian conquest of Babylon (Daniel 6:28).

At the decree from Medo-Persian kings Darius, Cyrus, and Artaxerxes, two groups of Jews returned from Babylon to rebuild Jerusalem and the temple. The priest Ezra led the first group at the edict of Cyrus, King of Persia. Nehemiah, cupbearer for King Artaxerxes, led the second group from exile back to Jerusalem. Despite much opposition from the Samaritans of the region, both the wall and temple were completed under the leadership of Governor Zerubbabel, High Priest Joshua, and Nehemiah.

Rebuilding of the Temple in Jerusalem

In the face of this resistance, prophets Haggai and Zechariah encouraged Zerubbabel and Joshua to complete the temple restoration. Here can be seen a connection between the temple reconstruction and God's plan to complete His eternal kingdom. To announce His purpose, Jesus responded to Jewish leaders who asked a sign,

> Jesus answered and said unto them, Destroy this temple, and in three days I will raise it up. (John 2:19)

Rebuilding the temple in the days of Ezra represented only the beginning. Starting in 20 BCE, King Herod the Great, as a concession to the Jewish leadership spent forty-six years dismantling and reassembling the temple. Though the work was done at great personal expense, Herod intended not only to restore its glory to the splendor of Solomon's first temple, but he desired to have the Jews acknowledge him to be their rightful king.

As an Idumean (part Edomite), Herod the Great did not have the proper lineage to be King of the Jews. The priests, suspicious of Herod's intentions, sought to protect the rolls of genealogy in order to prevent Herod from destroying them during reconstruction.

Nehemiah Restoring the Walls of Jerusalem

The Lord spoke to Zechariah and told him to prophesy to the High Priest Joshua that there would come forth one called the BRANCH, and He would build a temple. He also instructed that crowns should be made for Him to rule. Historically the office of high priest did not also include the dual role of king. Zechariah declared,

> And they that are far off shall come and build in the temple of the Lord, and ye shall know that the Lord of hosts hath sent me unto you. And this shall come to pass, if you will diligently obey the voice of the Lord your God. (Zechariah 6:15)

In Zechariah (4:1-4, 4:11-14) the two anointed ones are revealed as olive trees or branches emptying oil into the seven lamps of the menorah. By comparison, the larger Hannukah menorah contains nine lamps. The book of Revelation (11:1-4) describes the two anointed ones as the two witnesses sent forth during the final half (1,260 days) of the seven-year *shabua* to prophesy against the rebellious nations. Following their prophesy the Antichrist overcomes and kills them, after which they are resurrected.

Joshua (Yahweh is salvation) and Zerubbabel (Seed of Babylon) are often described as the two anointed ones. Both Joshua and Zerubbabel reflect the ministry of Jesus Christ as High Priest and as King. Another observation based on the name Zerubbabel arises due to the fact that he had his birth in a fallen kingdom. Jesus came into the world, born of a virgin that He might be God with us (Hebrew - *Immanuel*). Because neither Enoch nor Elijah died a natural death, they have also been suggested as the two witnesses who are killed and resurrected (Revelation 11:3-12).

The prophet Malachi, in the final book of the Old Testament, prophesied of the coming of the Lord's messenger,

> Behold, I will send my messenger, and he shall prepare the way before me: and the Lord, whom ye seek, shall suddenly come to his temple, even the messenger of the covenant, whom ye delight in: behold, he shall come, saith the Lord of hosts. (Malachi 3:1)

Malachi also spoke of the coming of Elijah,

> Behold, I will send you Elijah the prophet before the coming of the great and dreadful day of the Lord: And he shall turn the heart of the fathers to the children, and the heart of the children to their fathers, lest I come and smite the earth with a curse. (Malachi 4:5,6)

The Book of Mark opens referring to the messenger as John the Baptist (Mark 1:2,3), who came in the spirit and power of Elijah. Jesus spoke to his disciples of John as the messenger sent to prepare the way of the Lord (Matthew 11:10).

Jewish Tribulation in Last Days

Prior to his death, Moses spoke in the Book of Deuteronomy as he stood on the threshold of the Jordan crossing (Deuteronomy 4:29-36). There he delivered final warnings and a rallying cry to the nation of Israel as they prepared to enter the promised land. He rehearsed the covenant promises from Yahweh with connected blessings and curses. As if to prophesy of their coming disobedience, Moses spoke of inevitable tribulation in the latter days.

> I call heaven and earth to witness against you this day, that ye shall soon utterly perish from off the land whereunto ye go over Jordan to possess it; ye shall not prolong your days upon it, but shall utterly be destroyed. And the Lord shall scatter you among the nations, and ye shall be left few in number among the heathen, whither the Lord shall lead you. And there ye shall serve gods, the work of men's hands, wood and stone, which neither see, nor hear, nor eat, nor smell. But if from thence thou shalt seek the Lord thy God, thou shalt find him, if thou seek him with all thy heart and with all thy soul. When thou art in tribulation, and all these things are come upon thee, even in the latter days, if thou turn to the Lord thy God, and shalt be obedient unto his voice; (for the Lord thy God is a merciful God;) he will not forsake thee, neither destroy thee, nor forget the covenant of thy fathers which he sware unto them. (Deuteronomy 4:26-31)

All of history and the Biblical record testify to the importance of the Jewish people and the nation of Israel in relation to God's closing drama of His Kingdom. The final scene of the battle hinges on the outcome of a remnant of one-hundred-forty-four thousand Jews who escape to the wilderness of Moab after Antichrist institutes the Abomination of Desolation and great tribulation.

Knowing his ultimate fate, Satan strives to prevent the remnant from being redeemed by Jesus Christ as long as possible. Satan's end game remains one of demanding acknowledgement of himself as the object of worship. He did not succeed during the temptation in the wilderness when he offered to relinquish the kingdoms of the world if Jesus would bow down and worship him. If Jesus had complied, it would have spared Him the agony of the cross.

The event on the final fulfillment of the Day of Repentance (Yom Kippur) represents a completion of the Kingdom of God. All seven Feasts of the Lord (Leviticus 23) have a yearly celebration, as well as a day of final fulfillment on the calendar day of that festival at a year to be determined. This condition was satisfied for the first four feasts of Passover, Unleavened Bread, First Fruits, and Pentecost in the year 33 CE. The same condition will be fulfilled for the remaining three; Trumpets (Rosh Hashanah), Day of Repentance (Yom Kippur), and Tabernacles (Succot).

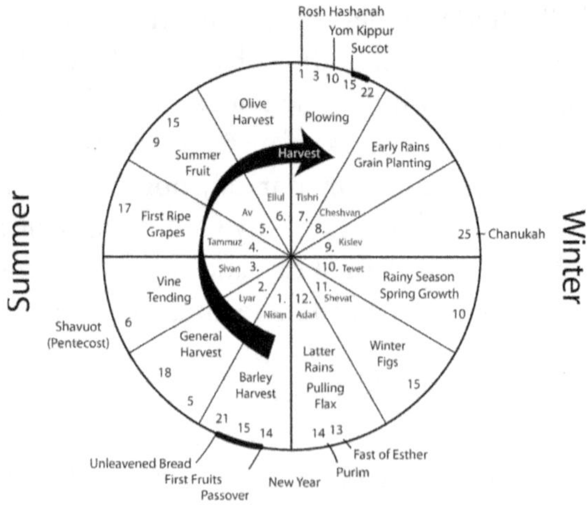

Circular Chart of Jewish Yearly Calendar

As with the first four festivals, the remaining three would be expected to take place on the same year were it not for a qualifying statement by Jesus. He declared regarding the duration of the Great Tribulation,

> For then shall be great tribulation, such as was not since the beginning of the world to this time, no, nor ever shall be. And except those days should be shortened, there should no flesh be saved: but for the elect's sake those days shall be shortened. (Matthew 24:21,22)

Based on the timing stated for the fifth trumpet (five months) and sixth trumpet (over thirteen months), the final Feast of Trumpets (Rosh Hashanah) must take place more than eighteen months earlier than the final two feasts (see Revelation 9:10-15). Given the timing of Rosh Hashanah, that means the catching away of believers, both alive and from the grave, must occur about two years before the year of the final fulfillment of Yom Kippur and Succot.

The Abomination of Desolation by Antiochus-IV Epiphanes

As explained in detail in my previous book *Rend the Heavens*, the catching away of believers recorded in Matthew 24, Mark 13, and Luke 21 coincides with signs on earth and in the heavens. The same dramatic signs are described in Revelation 6:12-17 at the opening of the sixth seal. This shaking of earth and parting of heaven represents the fulfillment of the Feast of Trumpets less than eighteen months after the placement of the Abomination of Desolation by the Antichrist.

Commencement of the seven-year *shabua* (Hebrew meaning seven or completion) will begin with an agreement patterned after the covenant at Beersheba by Abraham and Abimelech. Beersheba in Hebrew means well of the oath, or well of seven. Abraham brought forth seven ewe lambs as a sign of the agreement of peace with Abimelech. Recall also, Jotham the son of Gideon prophesied that a similar agreement between Abimelech the illegitimate son of Gideon, and the men of Shechem would result in mutual destruction (Judges 9:19,20).

In the plan of God, the remnant of Jews will be hidden in Petra (or Sela) in the wilderness (Revelation 12:14-17) until Jesus Christ appears to them in person (Revelation 14:1-5) to show the wounds in his hands,

> And one shall say unto him, What are these wounds in thine hands? Then he shall answer, Those with which I was wounded in the house of my friends. (Zechariah 13:6)

After the appearance of Jesus Christ to the one-hundred-forty-four thousand Jews and their acceptance of Him as the Christ, they will go to confront the Antichrist before the seven bowls of the wrath of God pour out on the earth (Revelation 15 and 16).

The Harlot in Purple and Scarlet Upon a Beast with Seven Heads

In the Book of Daniel, a vision appeared to him as of a beast with seven heads and ten horns (Daniel 7:7,8,24). Upon the beast a small horn came up having the eyes of a man and displaced three of the horns. The little horn spoke with a mouth saying great things. Daniel's imagery matches that of the dragon (Revelation 12:3,17:3,7-13) and the scarlet-colored beast with a harlot dressed in purple and scarlet. These represent the powers of darkness allied with the apostate church. Ironically, the beast with seven heads and ten horns represent worldly kings who hate the harlot and devour her.

In the symbol of Nebuchadnezzar's dream, these are the ten toes of the great image that rise in the last days. When they make war with the Lamb, He overcomes and destroys them (Revelation 17:14). These are broken by the stone coming down from heaven to strike the feet at the return of Jesus Christ (Daniel 2:34,35).

[illegible]

[illegible]

Illustration of Open Bible

Chapter Fourteen - Read the Script

For anyone not experienced in a very complex skill, it is difficult to imagine stepping out on short notice and performing flawlessly from memory. Pianist Maria Joao Pires received a phone request in 1999 to substitute at the last minute at the Concertgebouw Orchestra in Amsterdam playing Mozart's Piano Concerto No. 20 in D-minor (K. 466). She mistakenly wrote down (K. 488) which is No. 23, the A-major Concerto, and practiced that piece accordingly.

When conductor Riccardo Chailly commenced Mozart's Concerto No. 20, Maria Joao Pires suddenly panicked, quietly vocalizing her error to the conductor. After a confident reassurance from conductor Chailly, Maria calmed herself, began the piano refrain, and performed the remainer of the concerto flawlessly. Pires was able to perform without practice because she had already internalized the concerto.

That incident displays the importance of reading and knowing the music in order to perform on command. For an actor, the script defines the performance. The ability of an actor or actress to put on the role and bring the character to life despite obstacles often determines success or failure of a production.

Coming from an accomplished musical family, I was expected to take up an instrument like all of my siblings. Unfortunately, the music teacher at school convinced my mother he needed a French horn player for the marching band. By the end of sixth grade, I refused to lug the oversized case, received a pardon, and was relieved of duty. If I had been allowed to take up the violin, it would have been a different story, but the marching band did not need a violin.

During His Olivet discourse, Jesus said,

> And this gospel of the kingdom shall be preached in all the world for a witness unto all nations; and then shall the end come. (Matthew 24: 14)

Though not every person has received a clear presentation of the gospel message, the Word of Life has reached all nations and tribes. Preaching the gospel message and making disciples represent the fundamental calling of believers. Many individuals testify to having come to faith by hearing a broadcast over the internet, even by a dream or direct vision of Jesus. By 2020, every person has had access to the message by one means or another.

Today, availability of the Bible in numerous versions and languages represents a modern-day miracle, on par with the fifteenth-century printing of the Gutenberg Bible. Despite the challenge of understanding the scope of Biblical knowledge, many language tools and study guides exist to satisfy any interested reader. Online Bibles apps negate any excuses for not knowing "The Greatest Story Ever Told".

Hebrew-Greek Study Bible

While in the process of preparing lessons for noon Bible studies thirty years ago at work, I came across a useful tool for Hebrew-Greek words. The reference Bible included important word meanings as well as a complete list of Strongs numbered words in both Hebrew or Greek. The Hebrew-Greek Bible version proved to be very helpful prior to availability of online reference aps.

It did not take long before I began to find a significant list of errors in the printing. In fact, I eventually detailed a full page of misspellings, misprints, and word substitutions in the Bible and reference sections. On several occasions the errors created confusion because the typographical mistake did not appear obvious.

The errors were surprising because the editor was a well-known Hebrew-Greek scholar. Rather than destroying the version, I prepared a letter to the Bible editor and publisher listing all of the errors. Neither editor nor publisher responded. However, several years later I chanced to see a new edition of that study Bible in a bookstore. A quick review of the pages indicated the errors had been corrected.

One of the most infamous Bible printing errors was termed the "Wicked Bible". In 1631 King Charles-I commissioned London royal printers Robert Barker and Martin Lucas to complete a printing of the King James Version. After 10,000 copies had been printed, two typesetting errors were discovered. In Exodus 20:14 the "not" had been omitted leaving, "Thou shalt commit adultery." The king fined the printer and ordered all copies destroyed. Less than twenty copies of the "Wicked Bible" remain in existence today.

Jewish scribes meticulously checked their hand written scrolls by counting characters. Any manuscript with more or less than the required number of characters was destroyed. For most English translations of the Bible, the total number of disputed verses can be summarized on a single page of paper. None of those verses reflect any substantial doctrinal issue.

Despite that, some Bible skeptics occasionally promote a shopping list of "so-called" contradictions. Most of those "contradictions" can be easily explained as nothing more than a misinterpretation. There are 25,000 full or partial early manuscripts of the New Testament, some written within a decade of the original. No historical document has been as well attested as the Bible with copies so close in date to original authorship.

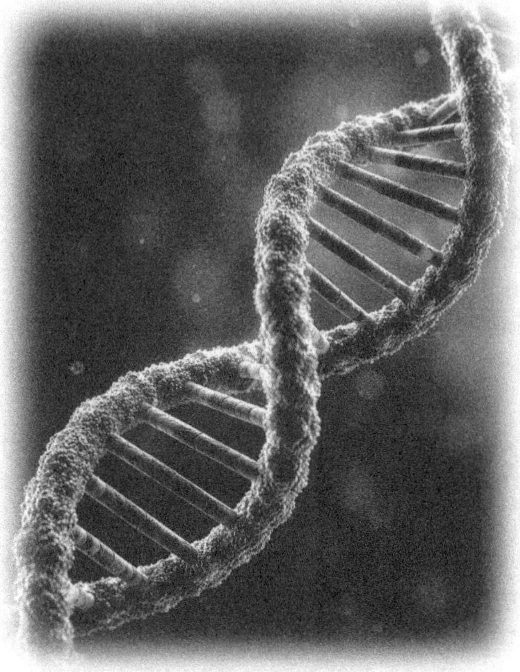

Information and DNA

Information Theory

During World War II, communication by analog-signal devices in the field and across oceans required regular amplification. Analog refers to an electromagnetic radio wave signal rather than a digital signal. Even when amplified, analog message quality diminishes due to introduction and intensification of background noise, resulting in information loss. In 1948, Bell Laboratories struggled to understand the problem and how to solve the challenge of reliable information transmission over long distances.

Enter brilliant, electrical-engineer Claude Shannon at Bell Laboratories. Shannon formulated a mathematical theory of information communication. Without going into the technical detail, Claude Shannon isolated the original message from the noise source by treating each as a probability distribution.

Coding a voice or data as digital bits rather than an analog signal allows the information to be "quantized." When transmitted as a digital signal, the message can be compressed to minimum signal bandwidth. Thus, a message can always be communicated reliably by autocorrelating against itself to isolate the received message and remove the noise. Auto-correction of the message assures accuracy of modern digital transmission and storage.

If I transmit the same message several times while introducing a random error each time, a listener can quickly decipher the original message. Each additional time the message is transmitted, the greater the likelihood the receiver will properly interpret the original message.

DNA (deoxyribonucleic acid), the building block of life, represents a classic example of digital information storage and transmission. From its digital nature of four, base-pair combinations, DNA allows for accurate replication of all the traits and machinery of a living thing or being.

In digital form replicated nucleotides can be checked and auto-corrected using cell-repair mechanisms. Similar to the digital communication mentioned above, the process of DNA and cell replication is exceptionally reliable.

Pontius Pilate questioned Jesus with, "What is truth?" (John 18:38) Pilate's probe is an example of the classic challenge to a statement of truth. What do we know and how do we know a belief is true?

What did Jesus respond to Pilate's challenge? Being truth Himself, Jesus did not need to respond to Pilate's question. He provided the answer when He spoke to His disciple Thomas, "I am the way, the truth, and the life: no man cometh unto the Father but by me" (John 14:6).

Jester at the Gaming Table

Chapter Fifteen - Know Your Lines

WHILE SPEAKING WITH family members and friends before a funeral, I started a conversation with an acquaintance who recently began attending a well-known Bible college. As we spoke, he suddenly exclaimed, "You are one of those people who believes the God of the Old Testament and God of the New Testament are the same person! Don't you realize the God of the Old Testament was angry and vindictive, but the God of the New Testament is merciful?"

Even in relatively conservative Christian colleges, that idea has become commonplace. A superficial reading of the Bible might give the impression God matured or changed His tactics between the Old and New Testaments. Only the Holy Spirit can truly enlighten the understanding to realize the context. That requires discernment.

The middle of the nineteenth century reflected a time of growing skepticism regarding the accuracy of Biblical teaching. During the period, many theological seminaries and secular universities fostered enlightened thinking that questioned divine inspiration as the source of

Scripture. "Higher criticism" represented an effort to extract meaning through reason and history apart from the traditional understanding of the Bible as the inspired Word of God.

Diverging Paths

In the mid-1940s, two evangelists came together focused on world evangelism. Charles Templeton (1915-2001) began his career doing a daily sports cartoon series for the Toronto Globe newspaper. After accepting Christ in 1936, Templeton began preaching and evangelizing throughout Canada and the United States. When he met Billy Graham at a crusade in 1945, the media regarded Charles Templeton as the more promising and articulate evangelist of the two. At the time, Billy Graham (1918-2018) had only recently begun his ministry.

Templeton and Graham worked together for several years speaking at crusades for the new organization Youth for Christ. During the early 1940s, Templeton experienced a crisis of belief regarding Scripture and the basis for faith. At a 1948 Los Angeles conference with Billy Graham, Charles Templeton challenged Graham to think critically about whether the stories of Scripture were true.

After their meeting Graham resolved that the Spirit of God brought results through his own preaching only when he proclaimed, "The Bible says…". Billy Graham concluded he would not struggle intellectually about the inerrancy of Scripture.

On the other hand, Charles Templeton chose to attend Princeton Theological Seminary in 1948 to get answers for his many doubts. After graduating from his studies at the liberal institution of Princeton, Templeton in 1957 decided to leave the ministry to return to a career in media and broadcasting. Charles Templeton published a book *Farewell to God* in 1996, detailing his departure from the faith.

In 1998, Charles Templeton sat for an interview with Lee Strobel prior to the publication of Strobel's book *The Case for Faith*. Near the conclusion of his meeting with Templeton, Lee Strobel posed a probing question. Asked by Strobel what he thought of Jesus Christ at this stage of his life, Templeton responded, "He was the greatest human being who has ever lived. He was a moral genius. …His commitment was total and led to his own death, much to the detriment of the world." Lee Strobel commented, "You sound like you really care about him." Charles Templeton replied, "Well, yes, he's the most important thing in my life." He concluded his evaluation tearfully saying, "I ...miss... him!" Once he regained control of his emotions, Templeton waved his hand dismissively, "Enough of that."

Just as a Lion, The Word of God Defends Itself

Running counter to the trend, Charles Spurgeon (1934-1892) sought to bring the church back to the foundational teaching of the Bible as the authoritative source of truth for life and practice. Spurgeon, an English Baptist preacher, was asked in 1860 how to best defend Biblical doctrine against the higher criticism of skeptics. In one of his most famous metaphors, Spurgeon likened the Word of God to a caged lion. He said,

> Defend the Bible? I would just as soon defend a lion! Open the door and let the lion out; he will take care of himself.

With that same sentiment, Billy Graham responded to Charles

Templeton that his own preaching had power and conviction when he declared, "The Bible says...".

Pre-incarnate Appearances of Jesus Christ

Does Jesus Christ also appear in the Old Testament? Scripture records specific appearances in the Old Testament where Jesus Christ manifested Himself in human or angelic form. Pre-incarnate describes a circumstance where the Son of God enters history and interacts with persons, often appearing and then departing with no further evidence. Jesus Himself declared, "...Before Abraham was, I AM." (John 8:58)

Such was the situation when three "men" arrived at the doorway of Abraham and Sarah's tent in Mamre (Genesis 18:1-22). In the interaction, Abraham addressed one of them as "My Lord". The "man" proceeded to promise He would return that Sarah might have a son, according to the time of life (nine months).

Upon departure of the three, Abraham detained the Lord while the two angels descended from the plain of Mamre to Jordan and the City of Sodom. In the ensuing negotiation, Abraham pled with the Lord to save the cities near Sodom and Gomorrah. Recall that Abraham's nephew Lot and his family now resided within Sodom. After bargaining with God, Abraham negotiated the deal from fifty down to ten righteous persons. Ten were not found.

Similar pre-incarnate appearances occurred when Jacob wrestled with the angel (Genesis 32:24-32) and when the angel promised Manoah and his wife a child Samson (Judges 13:15-23). Manoah asked his name whereupon the angel replied, "Why askest thou thus after my name, seeing it is secret?" Hebrew *Pali* means secret or wonderful, just as prophesied by Isaiah regarding the Messiah,

> For unto us a child is born, unto us a son is given: and the government shall be upon his shoulder: and his name shall be called Wonderful, Counsellor, The might God, The everlasting Father, The Prince of Peace. (Isaiah 9:6)

Old Testament Visions of Jesus Christ

A vision differed from an appearance in that the person sees an image of Jesus Christ whether on earth or in the heavens. Similar events happened to Moses (Exodus 24:9-18), to Isaiah (6:1-8), to Ezekiel (1:1-28), to Daniel (12:1-13), as well as many others. Though circumstances and appearances varied, in each case the Son of God appeared in a heavenly vision to reveal Himself.

Prophesy of Jesus Christ

There are at least three hundred significant prophetic Scriptures fulfilled by Jesus Christ. They cover His calling, birth, ministry, miracles, death, resurrection, and His ascension. As of now, the remaining prophesies yet unfulfilled are those related to the return of Jesus Christ at the catching away, His coming to rule and reign, and His rule in eternal glory. It is statistically impossible for any one person to have fulfilled even thirty of those 300 prophetic Scriptures.

Metaphors of Jesus Christ

Following are several of many references to the person and work of Jesus Christ by metaphor. Use of a metaphor creates a picture of Jesus. These are important because Old Testament descriptions foreshadow His nature and character as well as His work.

God promised to Eve a SEED that would crush the head of the serpent (Genesis 3:15). The SEED also appears in the name of the governor Zerubbabel (seed born in chaos) who with High Priest Joshua rebuilt the temple (Ezra 3:8).

Noah (Hebrew – rest) received a call from God to build an ark which included a DOOR (Genesis 6:16). Noah's family and selected animal enter the DOOR to be saved from judgment. Jesus also described Himself as the DOOR (John 10:7-9).

When Abraham returned with the spoils from overthrow of the gentile nations (Genesis 14:18-20), he gave a tithe to Melchizedek (King of righteousness) whom Scripture describes as King of Salem

(Hebrew – peace), an early name for Jerusalem. Melchizedek is also identified as a priest of the most high God.

The prophet Zechariah prophesied of God calling forth His servant the BRANCH (Zechariah 3:8) who springs forth to fulfil the Kingdom of God and build His temple (Zechariah 6:12).

The Name of Jesus Christ

The name Jesus the Christ (Hebrew – *Yeshua Hamashiach*) has the literal meaning "YHWH is salvation, anointed is He". In Psalm 28, a literal translation of the verse says,

> YHWH is strength for them and a stronghold of *salvation anointed is He.* (Psalm 28:8)

The final three words in Hebrew declare, *Yeshua Hamashiah Hi.* In other words, He is Jesus Christ. Someone might argue, "The words are not the name." Yet Jesus Christ is the Word. Meditate on that if it is not clear; the Word is the name and He is the Word.

A similar expression in Psalm 20 declares literally,

> *Salvation brought YHWH for His anointed* He will answer him from the heavens of His holiness might with the saving of His right hand. (Psalm 20:6)

The statement opens with *Hoshea YHWH Hamashiah.* Again, the Word is the name and He is the Word. These psalms record expressions of King David (Hebrew – beloved), a pre-figure of the person Jesus Christ, the Son of David.

Isaiah (Hebrew – salvation is YAH) spoke of the coming anointed One who is sent forth from the beginning by God and His Spirit, saying,

> Hearken unto me, O Jacob and Israel, my called; I am he; I am the first, I also am the last. ...I, even I, have spoken; yea, I have called him: I have brought him, and he shall make his way prosperous. Come ye near unto me, hear ye this; I have not spoken in secret from the beginning; from the time that it was,

there am I: and now the Lord God, and his Spirit, hath sent me. (Isaiah 48:12,15,16)

This verse embodies the clearest presentation of the triune personhood of God and His plan to enter human history.

Moses Pointing to A Brass Serpent on a Pole

Serpent in the Wilderness

Part of the legacy resulting from forty years in the wilderness included a pole with a brass serpent atop it (Numbers 21:4-9). It began with the people of Israel murmuring about the quality and limited choice of manna (bread), as well as lack of water in the desert. In response the Lord sent poisonous snakes (fiery serpents) among them which bit many fatally.

When the people repented, God gave instructions to Moses for the creation of the pole with a brass (or bronze) image of a poisonous snake on it. Looking on the pole and image was sufficient to heal anyone bitten and prevent death. The brazen serpent had a useful purpose, at least for a time, though it meant much more.

The people of Israel carried the resulting relic into the land of Canaan where it remained among the instruments maintained by the priests. Five hundred years later, during the reign of King Hezekiah (715-686 BCE), the brazen serpent had been elevated from a relic to an object of worship. As part of Hezekiah's rule, he instituted reforms to clean out the Temple in Jerusalem, destroy idolatry, and establish proper worship again.

When Hezekiah identified the brass serpent and the way it had been used as an article of veneration (2 Kings 18:4), he had it broken to pieces, rightly referring to it as *Nehushtan* (a piece of brass). When speaking to the Jewish leader Nicodemus, Jesus noted the principle for the brazen serpent as a foreshadowing of His purpose.

> And as Moses lifted up the serpent in the wilderness, even so must the Son of man be lifted up: That whosoever believeth in him should not perish, but have eternal life. For God so loved the world, that he gave his only begotten Son, that whosoever believeth in him should not perish but have everlasting life. For God sent not his Son into the world to condemn the world; but that the world through him might be saved. (John 3:14-17)

The image of their punishment not only healed the afflicted, but it looked forward to the ultimate fulfillment, the person and work of Jesus Christ on the cross. He took the consequence of all sin upon himself. That insight does not result by focusing on the symbol but rather recognizing and believing the substance behind it.

Mark Twain's Book, *The Innocents Abroad*

In conjunction with the 1867 Paris Exhibition, a group of American tourists embarked on an excursion from America to Europe and the Middle East including the Holy Land. Writer Mark Twain (pen-name for Samuel Clemens, 1835-1910) accompanied the tour under contract to write articles about his travels. In 1869, Mark Twain republished the essays in the form of a travel guide titled *The Innocents Abroad*, written for the general public based on his excursion. Twain's writing exposed the folly of certain religious sentiments.

Composed as humorous satire, Mark Twain documented the naïve tourist mindset of his religious companions. In particular, during travels of the Holy Land, he observed that many of the travelers defaced historical or holy sites in order to collect souvenirs, the stereotypical relic-hunters.

Mark Twain likewise spared no mercies on the European and Palestinian tourist industry of the day. Tour guides and shopkeepers abounded with fanciful artifacts and stories crafted from a grain of truth to entertain gullible travelers. Many of those fables still provide filler material for undiscerning sermons today. The "eye of the needle" gate story serves as an example.

The False Eye of the Needle Gate Story

While teaching a Sunday School class, I related the oft repeated illustration about a small gate in Jerusalem's wall referred to as the Eye of the Needle. Jesus used the phrase in reference to the impossibility of a "rich man" getting into heaven based on riches compared to a camel going through the eye of a needle (Matthew 19:23-26). As the story goes, a camel would need to be unburdened to fit through the small gate. The moral of the story would be that a rich person must be "unburdened" to enter the Kingdom of Heaven.

But after referencing the illustration in class, I clarified that the "so-called" Jerusalem gate was a fabrication. There never existed such a gate in the wall of Jerusalem. The narrative happens to be a medieval illustration, popularized by Thomas Aquinas, but not associated with any real place or event. Though an interesting fiction, the eye of the needle story does not add clarity to the lesson told by Jesus.

When I stated that the gate was fictitious, several persons in the class expressed they were upset because they heard the story from a tour guide during a Holy Land trip. As Jesus explained, it is impossible for riches to provide access to heaven. The hearers never took the time and discernment to verify whether the Jerusalem gate story is actually true. To be an intelligent listener or reader requires critical thinking skills.

Imagine This

Shortly after the breakup of the Beatles, John Lennon wrote and released a song entitled *Imagine* (1971). Woven through the lyrics are a wish list of undesirables to be imagined away: "no heaven", "no hell", "no religion", "no countries", and "no possessions". On the flip side of his hopeful coin Lennon imagines "peace" and "brotherhood" and "sharing".

The great contradiction of the contrast constitutes what can be called, the *tyranny of unintended consequences*. The quest to achieve Lennon's "world as one" means something or someone must die.

In Chapter 4, conflicting stories of the American Revolution and the French Reign of Terror demonstrate the deception of focusing on outcomes. During the Reign of Terror, a populist version of the French Revolution motto expressed three outcomes and the alternative: *Liberty, Equality, Fraternity, ...or Death*. In practice the implementation instead reflected unintended outcomes: Anarchy, Parity, Sodality, ...or Murder.

Sans-Culottes Common Clothing of the French Revolution

Elimination of rule (anarchy) yields the tyranny where everyone rules. Equal result rather than equal opportunity imposes parity by confiscation. The idea of fraternity means willing association whereas the Latin word sodality implies compulsory comradeship.

The story of *L'Acedia, Land of Leisure* in Chapter 18 mirrors the degeneration of the social fabric by unintended consequences. As with Nimrod and Babylon, when fallen man and angels intervene to control outcomes, chaos happens from Genesis (10:8-10) to Revelation (17:1-18).

The Modern Liberal

The apparent concern of Judas for the poor certainly displayed his liberality. What does it mean to be liberal? Is liberality merely a contrast to conservative? Notice the first warning of Jesus on the Mount of Olives. When teaching of His return He cautioned, "Take heed that no man deceive you." (Matthew 24:4) In an effort to clarify terms, the modern meaning of liberality in educational, political, economic, or social issues does not reflect the biblical position. Isaiah prophesied:

> The vile person shall be no more called liberal, nor the churl [an impolite or mean-spirited person] said to be bountiful. For the vile person will speak villainy, and his heart will work iniquity, to practice hypocrisy, and to utter error against the Lord, to make empty the soul of the hungry, and he will cause the drink of the thirsty to fail. The instruments also of the churl are evil: he deviseth wicked devices to destroy the poor with lying words, even when the needy speaketh right. But the liberal deviseth liberal things; and by liberal things shall he stand." (Isaiah 32:5-8)

True, biblical liberality refers to someone with a generous and selfless spirit. The perfect example of liberality was Jesus at the Last Supper, offering himself for the good of the world and telling his disciples, "...do this in remembrance of me." (Luke 22:19) The emphasis here should be placed on "do" more than on "remembrance". That is, do what Jesus did to give of Himself with a generous and selfless spirit.

Contrast the Bible's liberality with Isaiah's churl who is "a rude, stingy, or morose person." Don't mistake true, scriptural liberality with the modern culture demonstrating churlishness as liberality. Under the inspiration of the Holy Spirit, Paul the Apostle properly expressed what people would be like in these times:

> This know also, that in the last days perilous times shall come. For men shall be lovers of their own selves, covetous, boasters, proud, blasphemers, disobedient to parents, unthankful, unholy, without natural affection, trucebreakers, false accusers, incontinent, fierce, despisers of those that are good, traitors, heady, highminded, lovers of pleasures more than lovers of God; having a form of godliness, but denying the power thereof: from such turn away (2 Timothy 3:1-5).

A False Dream

Shortly after publication of my recent book *Rend the Heavens*, a very close relative passed away after a long illness. For eighteen months before his passing, an acquaintance named Tim helped and ministered to him. A few weeks before his passing, I visited my relative in a nursing facility as he began hospice care. He thanked me for giving him a copy of my book.

During our conversation the next day, he posed several questions regarding the book. After answering my relative's questions, I invited him to place his faith in Christ, whereupon he responded, "No, and I'll tell you why I can't do that now." He proceeded to explain a vivid dream he experienced recently.

In the dream he stood before the throne of God who supposedly told him, "I don't know what I am going to do with you. You are so unique I can't send you to Hell or the Devil will leave, and I can't let you into Heaven. I have no place for you." After hearing his story, I explained that the dream was deceptive. God does not face any dilemma, nor does He struggle with choices. Outcomes are clear because God, through His Son Jesus Christ put the burden of choice on each of us. To that he responded, "That dream is my reality. Don't tell me my reality is not true."

The Strait and Wide Gates

Jesus described only two gates; the strait (narrow) gate He supplied, or a wide gate representing every other option. As Jesus stated,

> Enter ye in at the strait gate: for wide is the gate, and broad is the way, that leadeth to destruction, and many there be which go in thereat: Because strait is the gate, and narrow is the way, which leadeth unto life, and few there be that find it. (Matthew 7:13,14)

Beliefs and experiences do not necessarily reflect reality; that determination requires discernment, which must be enlightened by the Word of God and His Holy Spirit. As Jesus declared,

> And Jesus answered and said unto them, Take heed that no man deceive you. For many shall come in my name, saying, I am Christ; and shall deceive many. (Matthew 24:4,5)

Shell Game

During a 2006 national church scouting event in Missouri, our local outpost participated in a campfire presentation. One of the leaders shared a story of his scouts during an earlier campout near St. Louis, Missouri. Given some free time before dinner, several of the older boys asked permission to walk to a nearby shopping area. Their spending money apparently began "burning a hole in their pocket". After receiving approval, the boys headed into town with instructions to stay together and return in one hour for dinner.

As they approached the store, a young man with a small table on the sidewalk called them over asking for help. They cautiously obliged and surrounded the table where the young man explained his dilemma trying to learn a new board game. As he began to explain the game, a couple of passersby took interest and joined. Anyone with street smarts knows the shell game, how it is played, and the predictable outcome.

With a quick demonstration the young gamesman let the boys try their luck guessing for free. Not surprising, they won. But because they bet nothing, they won nothing. Since the gamesman claimed to be learning himself, he generously gave them a second try for free. Again, they won but received nothing. Now the two spectators began to congratulate the boys profusely on their luck, challenging them to put a few dollars on their next guess.

To everyone's amazement the boys won again and doubled their money. They were hooked and needed little persuasion to place a bet again in addition to adding their prior winnings. The winning streak continued as they each received a small jackpot. By now caution set in and the boys agreed to call it quits while they were ahead.

In a chorus of protest the gamesman and spectators shamed them into trying again with all the money in their pocket. Reluctantly they emptied their pockets, confident they would not lose; but they did. Suddenly the game ended in a rousing disturbance and everyone scattered. Despite their protests, the boys were left with empty pockets and bewildered. What just happened they wondered among themselves?

By the time the boys arrived back at camp they had already reasoned among themselves the trap that transpired. With their attention distracted by easy wins and a couple of shills urging them on, they easily fell prey to sleight of hand. At no point did they have control of the game, nor did they understand until it ended. Their scouting leader related the hard lesson and how easily they had been drawn in.

As I was teaching a Sunday School class, one of the ladies related that her husband had a compulsive gambling habit. She explained every time he won big at the tables; he could not sleep for days until he returned to the tables and lost everything.

During more than a decade of jail ministry, I have heard many similar stories of men who had a good job, a nice family, and a beautiful home. Before they knew it, everything was lost to a bad decision.

The circumstances varied and the sleight of hand differed but the outcome never changed. When playing against the house, you lose even when it appears you are winning. Satan has many distractions and a host of shills like the jester who are proficient at the game. With the approaching completion of the Kingdom of God, Jesus forewarned that Satan would deceive many.

> For there shall arise false Christs, and false prophets; and shall shew great signs and wonders insomuch that, if it were possible, they shall deceive the very elect. Behold, I have told you before. (Matthew 24:24,25)

Ticket Booth at Theater Entrance

Chapter Sixteen - Places Everyone

IN 1939, PROMISING child actor Shirley Temple received top billing making her the obvious choice to play Dorothy Gale in the movie *The Wizard of Oz.* At age eleven, Shirley closely matched the age of Dorothy in the book version. MGM (Metro-Goldwyn-Mayer) sought to secure Shirley Temple for the movie, but at the time she had a binding contract with 20th Century Fox studios.

Two other actresses under consideration to play Dorothy were Judy Garland and Deanna Durbin, both age sixteen during casting. Though MGM selected Judy Garland based on her superior vocal range, she and Deanna Durbin would both need cosmetic and costume adjustments to make them appear younger.

Despite a very early and successful acting career, Judy Garland (1922-1969) experienced a range of personal problems as she grew older, including addiction and five failed marriages. She died from an overdose of sleeping pills, while out of work and impoverished.

Casting remains one of the entertainment industry's greatest challenges. A star actor or actress can promote an otherwise poor script. On the other hand, engaging top talent for the wrong part occasionally ruins an otherwise successful movie or acting career, especially when the actor has a reputation as "high maintenance". Often a prodigy star emerges based on an unknown actor appearing in a promising role.

Theater casting experiences greater challenges. Managing live theater casting has become infinitely more complex because of low budgets, scheduling challenges, and dependence on part-time actors.

A Star is Born

For screen, stage, or music, certain children demonstrate skills that identify them as a prodigy. What happens when you set out to develop a child prodigy?

William Sidis (1898-1944) had the greatest reported IQ, estimated to be as high as 300. His parents Boris and Sarah Sidis were two Russian Jewish professionals who influenced their son William to excel from a very early age.

As a Harvard educated psychologist, Boris Sidis played a leading role in William's development. He groomed his son at an early age so that he could read the New York Times by age 18 months. William was able to read and write eight languages by eight years of age.

By the time he turned eleven, William Sidis gained admission to Harvard University where he excelled academically. But extreme media attention and the inability to fit in with older students created resentment. After graduation from Harvard, William Sidis turned his back on learning and sought to live as a recluse, accepting only menial jobs. In 1919 Sidis participated in a socialist march and spent eighteen months in prison. He died in 1944 at age 46 after living most of his life in seclusion. The story of William Sidis and his career provide a classic example of a great plan derailed early.

Thankfully, not every train derails. The story of Edith Stern turned out quite the opposite. Aaron Stern experienced confinement in a German concentration camp during World War II where he became convinced stupidity fostered evil. After marrying Bella, they moved to Florida where their daughter Edith was born in 1952.

At a press conference, Aaron Stern announced he would make his daughter into a genius by raising "a perfect person". Due to a disability, Aaron Stern spent his time developing a process he called "total education immersion", and subjected Edith to a steady process of learning. He labeled the raising of his daughter "*The Edith Project*".

By age four, Edith read the complete *Encyclopedia Britannica*, and at twelve she entered college. She graduated Florida Atlantic University with a Bachelor's Degree in Mathematics at age fifteen and obtained a Master's Degree in Mathematics from Michigan State University. Edith Stern completed a very successful career as a Vice President at IBM with 120 patents to her name. Her father Aaron Stern eventually published the story of his project as *The Making of a Genius* (1971).

Professional Integrity

Peer review is used to achieve better quality publications put out by industry, academia, and government. What is peer review? Before a paper or project is published, the draft manuscript is sent to industry peers for them to review the methods used and conclusions reached. They may even try to replicate the original work to produce the same result.

However, peer review also has its dark side. After publishing a number of my peer reviewed technical papers on turbomachinery aerodynamics several decades ago, a leader of the fluids committee requested my help with a review. The publications to be reviewed involved technical papers at an upcoming fluid dynamics conference in Amsterdam.

This included the preparation for peer review of all technical documents that would be presented. The leader provided a list of conflicts and issues related to each author and reviewer. His objective in sending me these warnings was not to improve the peer review but to avoid personality skirmishes that would derail the paper-review process.

A more sinister problem in the technical arena involves a failure of professional ethics. Several outrageous instances involve recently retracted peer-review papers on research into professional ethical behavior.

The hypocrisy involved authors of ethics publications accused of autogenerating survey results, copying uncited text from other articles, and deductions inconsistent with data. Some of the biggest scandals resulted from reviewers and publishers who did not properly vet papers before publication, leading to retraction of the article.

Infernal Places

Author Dan Brown's book *Inferno* (2013) lays out a plot eerily similar to the recent 2020 COVID-19 pandemic. Rather than being prophetic, the book foreshadows much of the underlying social and political intrigue evident during the 2020 epidemic. Following the genre of his Professor Robert Langdon series (such as *The DaVinci Code*), Dan Brown framed *Inferno* as a techno-thriller conspiracy.

In the book, symbologist Professor Langdon uncovers a plot by billionaire genetic researcher Bertrand Zobrist to release a water-borne virus designed to sterilize a portion of those infected. Zobrist's expressed motive is to save civilization by reducing world population. As so often happens, evil cloaks itself in a robe of humanistic ambition and virtue.

In the novel *Inferno*, Dan Brown used Dante Alighieri's poem Inferno from *The Divine Comedy* (1321) as a backdrop with hints to unravel the mystery in his book. Throughout the story's plot, characters maneuver between roles of good and evil while trying to either discharge or prevent release of the virus. The dilemma thickens when Zobrist himself commits suicide. Despite efforts to the contrary, the virus is released.

A recurring quote by antagonist Bertrand Zobrist surfaces in Dan Brown's *Inferno*, "The darkest places in hell are reserved for those who maintain their neutrality in times of moral crisis." Though attributed to Dante, it actually reflects a variation on a statement by President John F. Kennedy as judgment on inaction during a predicament.

Every crisis, whether real, imagined, or manufactured, presents a dilemma. At the arrest, trial, and crucifixion of Jesus, everyone involved or observing those events missed the point. Only after His resurrection did certain followers and adversaries come to understand the redemptive purpose of God's plan.

Spiritual apathy or disinterest (Latin *acedia*) reveals the real plot of the enemy. As stated by G.K. Chesterton, "There are no uninteresting things, only uninterested people." Apathy fuels the Jester's conspiracy.

Joseph and Mary Find Jesus in the Temple

Chapter Seventeen - The Stage is Set

THE 1846 DONNER-REED wagon party experience described in Chapter Two tells a tragic tale. James Reed made the decision to take a shortcut called the "Hasting's Cutoff" to California. In their haste they followed bad advice and ignored warnings. The resulting delay caused them to spend the winter of 1846-47 snowbound in the high Sierra Nevada mountains. Almost half of the Donner-Reed party perished. The story represents a cautionary tale about shortcuts, particularly regarding end times and the return of Jesus Christ.

Two millennia ago, Jesus went through His early adult years on this earth. At age twelve Jesus remained in Jerusalem after the Passover (Luke 2:41-50), only to be found days later in the temple debating with the doctors of the Jewish law. Somewhere between the ages of twelve and thirty, His earthly father Joseph passed away, though details of his death are not recorded.

As oldest son of the household, Jesus would likely have been the primary family provider. Scripture records the public recognition of both Jesus and Joseph as carpenters (Matthew 13:55, Mark 6:3). The Old and New Testament words can be translated more accurately as craftsmen (Hebrew – *charashim*, Greek – *tekton*).

The Bible's first five books, called the Pentateuch, describe craftsmen as workers with wood, stone, metal, and cloth for fabricating tabernacle instruments. Throughout the Old Testament, craftsmen also did the work of assembling city walls, temple construction, and general building.

In Hebrew, the word *ben* identified a man as son of his father, though Jesus never referred to himself as *Yeshua ben Yoseph*. Rather, Jesus described himself occasionally as Son of man. Son (*Ben)* derives from Hebrew *banah* meaning to build or a builder of a house. Therefore, a man became builder of the house of his father; physical, generational, as well as spiritual.

Jesus, referring to His own body, said, "...Destroy this temple, and in three days I will raise it up." (John 2:19) In His death and resurrection, Jesus literally fulfilled His title as builder of His Father's house (Son of God), as well crafting an inheritance for all men (Son of man).

Based on the Biblical record, historical happenings, and astronomical events, the crucifixion of Jesus occurred on Passover, Friday April 3, 33 CE. Some researchers have instead placed the crucifixion on Passover the year 30 CE. However, that earlier date does not match the timing for the beginning of Jesus' ministry in addition to inclusion of three Passovers mentioned in the gospel of John (John 2:13-23, John 6:4, John 11:55-12:1).

The Prophetic Clock

Today we are on the threshold of two thousand years after the crucifixion with the world tottering on the brink of social upheaval. Throughout the preceding chapters, personal and historical accounts set the stage for God's majestic production in the last days. Jesus will return in the midst of great distress, much like we experience now (Daniel 12:1). The timing for His return rests in God's plan and purpose, but a coming peace agreement between Israel and her adversaries will set a prophetic clock in motion.

Abraham and Abimelech Make a Covenant of Seven

That clock refers to a seven-year (*shabua*) commencement marked by a covenant. The coming agreement will be patterned after the covenant between Abraham and Abimelech at Beersheba (well of the oath, or seven, or completion) when Abraham brought forth seven ewe lambs to seal the oath. At some point an Antichrist figure will confirm that covenant. The seven-year period (7 x 360 = 2520 days) ends on a literal fulfillment of the Day of Atonement (*Yom Kippur)* at which time the remnant of one-hundred-forty-four thousand Jews come to a face-to-face confrontation with Jesus Christ and acknowledge Him as their Messiah.

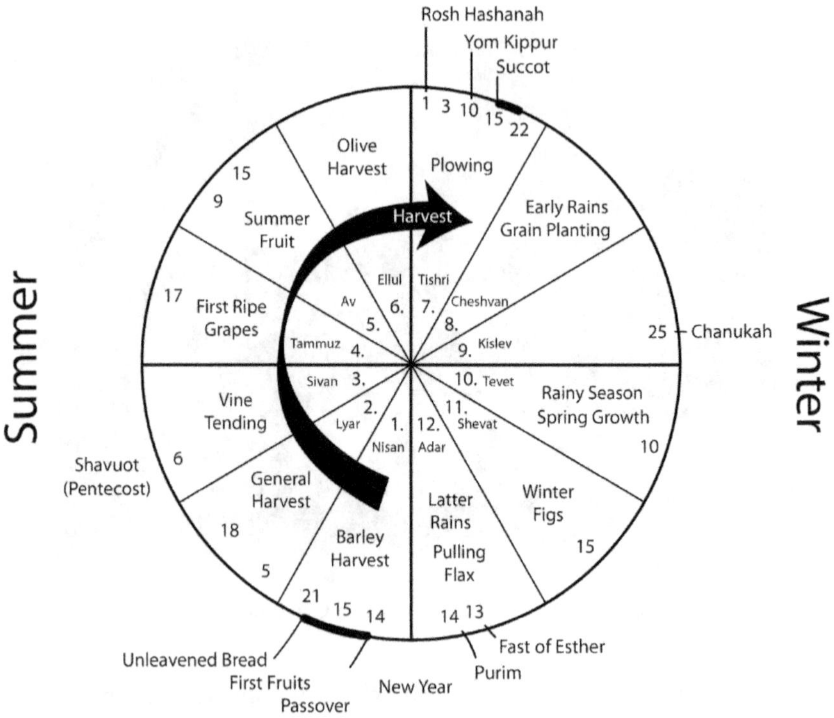

Jewish Prophetic Calendar

The Prophetic Calendar

A 360-day year in the Jewish prophetic calendar, must be reconciled to the 365.25-day astronomical calendar by adding days according to a predetermined schedule. The Jewish civil calendar employs a "luni-solar" system utilizing 29.5-day months with a leap month Adar added seven times every nineteen years.

Beginning with the signing of a covenant agreement, there are to be 1260 days until the raising up of an image in a tabernacle or temple. The statue called the "Abomination of Desolation" (Matthew 24:15, Daniel 9:27) will commence a worldwide attack on Christians and Jews, resulting in the martyrdom of many over the following eighteen months. Rather than progressing to the end of the second 1260 days (42 months, or three and a half years), Jesus declared the great tribulation would be cut short lest none should survive. The final Day of Atonement occurs at the end of the second 1260 days.

> For then shall be great tribulation, such as was not since the beginning of the world to this time, no, nor ever shall be. And except those days should be shortened, there should no flesh be saved: but for the elect's sake those days shall be shortened. (Matthew 24:21,22)

Throughout the past two millennia Christians, Jews, and even the population at large have faced tribulation. Many stories in preceding chapters describe great distress such as the 1930s great depression when the world faced severe economic and social turmoil. Adolf Hitler's pogrom labelled the "Final Solution" endeavored to exterminate Europe's Jewish population during World War II. It is estimated that six million Jews died as a result. These circumstances of persecution pale in comparison to what Jesus described as the Great Tribulation.

Historical evidence records about seventy million Christian that have been killed for their faith in the past two millennia, while over half took place in the twentieth century. Most of those cases of martyrdom occurred in the Soviet Union from 1918 to 1939 as well as Fascist and Communist conflicts in Europe from 1945 to 1953. During the Communist Chinese uprising beginning in 1949 many Christians were executed as well as in other Southeast Asian revolutions including North Korea, Laos, and Cambodia.

Year 2026 is seven years prior to the 2000-year anniversary of Jesus Christ's crucification. Commencement of the final seven-year *shabua* of Daniel's Seventy-Sevens hinges on a covenant agreement for Israel's protection, patterned after Abraham's Beersheba (well of seven, or completion) covenant with Abimelech.

In order for the seven-year *shabua* completion to fall on Yom Kippur (October 2-3, 2033), the covenant would need to be signed 2520 days earlier on November 8-9, 2026. Two days are denoted because in the Jewish and biblical calendar, a day begins at sundown and ends the following sundown.

Timing of the Abomination

As noted above, 1260 days after the covenant agreement initiates, there will be an abomination placed in a tabernacle or temple designated as the place of sacrifice. One step toward the sacrificial system renewal involves a qualified red heifer offering. Ashes of a kosher red heifer yield the ceremonial ashes for purification.

The modern Sanhedrin's official position declared the "trial" red heifer sacrifice in July 2025 as invalid for several reasons. But steps to establish the place and process for sacrificial renewal could be just months away. The sacrifice of Jesus Christ Himself suffices as the atonement for sin, yet restoration of tabernacle or temple atonement opens the door to an Antichrist figure claiming himself as a messiah.

What event falls 1260 days after the covenant? As stated above, if that agreement takes place on November 8-9, 2026, then the abomination should be placed on April 21-22, 2030. That is three and a half years later. Checking the calendar, Easter takes place on April 21, 2030. I prefer to describe that as Resurrection Day. Satan, through his minion Antichrist, would like nothing more than to "rise up" and declare himself as god and messiah on Resurrection Day. This is more than just a mere coincidence. Later in this chapter, a section on The Beast describes how an Antichrist figure rises up after a deadly wound.

When Jesus alluded to the destruction of the temple in Matthew 24:1,2, His disciples came to Him privately saying,

> Tell us when shall these things be? And what shall be the sign of thy coming, and of the end of the world? (Matthew 24:3).

Jesus began by warning of the risk for deception (Matthew 24:4,5). He then stated the popular pronouncement, "And ye shall hear of wars and rumours of wars..." (Matthew 24:6). Jesus followed the same verse with, "...see that ye be not troubled: for all these things must come to pass, but the end is not yet."

Wars and rumors of wars have occurred throughout history, but do not indicate the imminent return of Jesus. The corresponding passage in Luke 21 beginning "Then..." signals a clear transition, saying:

> Then said he unto them, Nation shall rise against nation, and

> kingdom against kingdom: and great earthquakes shall be in divers places, and famines, and pestilences; and fearful sights and great signs shall there be from heaven. (Luke 21:10, 11)

The Greek for nation is the word *ethos* representing ethnic divisions rather than geographic locations. No explanation can adequately portray the modern tide of tribalism along ethnic lines. The worldwide problem of ethnic conflict has torn apart families and nations. As discussed, no greater challenge exists than hearing the loudest voices call out a racism problem "out there" and failing to look in the mirror and to realize it is with me, "in here." A personal problem always forms the root of the general problem.

Beginning Birth Pangs

What is tribulation? Tribulation is not a seven-year period of distress! The Hebrew word *tsarah* for tribulation appears extensively in the Old Testament (Genesis 42:21) and means trouble or distress. Likewise, the Greek word *thlipsis* meaning tribulation reflects similar affliction. Jesus spoke in the synoptic gospel accounts of trials and persecution arising first from ethnic conflict, famines, pestilences, and earthquakes as the beginnings of sorrows (or birth pangs).

These stand as a forewarning of things to come (Matthew 24:8). Although similar conditions exist from place to place in the world and time to time through history, these events noted in the Olivet Discourse signal a dramatic, global intensification.

Tribulation has global and historical implications, far beyond the final seven-year period called in Hebrew *shabua* or seven or completion. Though suffering, distress, and a limited time of great tribulation occur within this seven-year covenant, it should never be falsely labeled "the tribulation."

Deception

Throughout Old Testament and New Testament periods, as well as world cultures, there is a tendency to embrace false teaching. Often deception occurs because of personal, professional, financial, social, or family pressures. Sadly, many Christians do not evaluate and make sound biblical judgments regarding doctrinal issues.

Jesus alluded to the matter in a sequence of verses leading up to the middle of the seven-year period (Matthew 24:9-14). Discernment requires good offense, knowing the objective and embracing the truth, as well as defense through understanding and confronting the assaults of the enemy.

Success of the gospel message fosters many alternative narratives, each with a different slant on three concepts of truth: the nature of God, the nature of man, and the means of reconciliation between God and man. Often differences in practice cloud the doctrinal scene and create unnecessary distinctives or denominations. Knowing the source of truth constitutes the best antidote to error.

Satan excels at mimicking, not originality. God creates all things as a reflection of His nature. Mankind creates as a reflection of God's nature and His creation. But Satan creates nothing; he only manipulates what God and man create to resist the purpose of God and man.

Roll with Seven Seals

Opening of the Seals

As with much prophetic revelation, the Apostle John stepped beyond the existence of space and time. Jesus Christ commanded John to, "Write the things which thou hast seen, and the things which are, and the things which shall be thereafter" (Revelation 1:19). After the Lord's instruction to seven churches of Asia, John beheld a door opened in heaven (Revelation 4:1-4). He heard a voice as a trumpet calling him to come up and be shown things to come.

In the spirit, John saw a throne and One that sat on the throne in heaven, in addition to twenty-four elders upon their seats and four beasts. Twelve heads of the tribes and twelve apostles of the church are already in God's presence. At the presentation of a scroll with seven seals and a Lamb slain, the beasts and elders appear with a multitude of angels worshiping the Lamb. In response, all creation in heaven, in earth, under the earth, and in the seas bless the One on the throne and the Lamb who alone is worthy to open the seals.

At this stage the Lamb commences to open the seals. Of seven seals on the roll (book) opened by Jesus Christ the Lamb, the story here focuses on the fifth and sixth seals. The timing for the commencement of opening the first seal is not recorded in relation to beginning the seven-year *shabua* period (84 months, 2520 days on the Jewish calendar based on a 360-day prophetic year). However, opening of the fifth seal certainly identifies the mid-point of the seven-years marking the placement of the Abomination of Desolation. Following are the seven seals.

1. White Horse – Rider going forth to conquer with bow and crown
2. Red Horse – Rider with power and sword to kill and take peace from earth
3. Black Horse – Rider with a balance bringing famine, ordered, "hurt not the oil and the wine"
4. Pale Horse – Death and Hell with power over a fourth of the earth to kill
5. Slain Saints – Saints slain for the Word of God given white robes pending the deaths of fellow servants

6. Cosmic Signs, Heavens Depart – Catching away and the harbinger of the coming Day of the Lord's Wrath. Sealing of one-hundred-forty-four thousand, Great Multitude in Heaven—Remnant in Sela sealed in their foreheads
7. Coming Seven Trumpets – Silence in heaven, the prayers of the saints, and the preparation of seven trumpet judgments followed by seven bowls of wrath

Of the seven, the first four seal judgments cause distress in the whole earth, with no outward singling out of believers. Though these trials appear to reflect intensifying affliction, they are not solely directed toward Christians and Jews.

The first seal reflects conquering by strength, the second by war, the third famine following war, the fourth death following famine. The fifth seal parallels the Jewish Holocaust carried out by Antiochus-IV Epiphanes at the Abomination setup in the temple as described in the Books of 1, 2 Maccabees.

These are also recorded in the twenty volume writings *The Antiquities of the Jews* by Flavius Josephus in 94 CE. The caution, "hurt not the oil and the wine" (Revelation 6:6) alludes to God's protection concerning the blessing experienced by patriarchs, prophets, and saints during a time of great dearth.

Death and Hell represented by the pale horse at the fourth seal denote power over a fourth part of the land. The wording does not necessarily imply a fourth of the population are to be killed, but rather death will stalk a quarter of the earth's territory to kill by sword, famine, disease, and beasts. Use of the Greek word Hades (meaning departed) for Hell comes from Greek *a-eido* or unseen.

The fifth seal, reflected in three verses (Revelation 6:9-11) describes the impending death of Christians and faithful Jews prior to the sixth seal cosmic events and the catching away of believers. Beginning with the Abomination of Desolation, a period of Great Tribulation addresses the wrath of Antichrist against the church and Jewish people. For that reason Jesus warned those in Judea to flee into the mountains (Matthew 24:16, Mark 13:14, Luke 21:21).

Events in Israel of the October 7, 2023 massacre by Hamas-led terrorists resulted in the death of 1,200 people including men, women, and children. Imagine a similar eighteen-months long campaign of Great Tribulation resulting in a worldwide massacre of Christians and Jews.

Of this period of months, Jesus said no flesh would survive were it not cut short prematurely (Matthew 24:22) by opening of the sixth seal. After the sealing of the one-hundred-forty-four thousand Jews and the appearance of the great multitude in heaven, the seventh seal opens the seven trumpets judgment of the Day of the Lord's Wrath.

As explained in this writing, a major reset of space and time and the natural order occurs at the sixth seal. At the rolling away of the fabric of the heavens, remaining believers and the dead in faith rise and are caught away before impending Day of the Lord's wrath. Explanation of the sixth seal cosmic events follows later in this chapter.

Adonikim and Adonijah

During the return from Israel's diaspora (scattering about) in Babylon and Persia, several groups were commissioned by Cyrus, Darius, and Artaxerxes to lead the Jews to Palestine (538-445 BCE). Both Ezra and Nehemiah recorded the names of key family leaders and their respective households by the number of their descendants. In the tradition of the Book of Numbers, God declared that Moses number the names:

> Take ye the sum of all the congregation of the children of Israel, after their families, by the house of their fathers, with the number of their names, every male by their polls; from twenty years old and upward, all that are able to go forth to war in Israel... (Numbers 1:2-3).

Both post-exilic writers Ezra and Nehemiah record the names of family patriarchs and their respective numbers saying:

> ...The number of the men of the people of Israel... (Ezra 2:2)

One individual among the names has significance based on the Hebrew meaning *Adonikam* which is "Lord of rising" or "Lord of resurrection." The name Adonikam portrays one who resurrects or rises up. The following lists of chief men with numbers of descendants are not inclusive of all names given in Ezra 2:3-35. However, the name Adonikam in each listing appears in essentially the same order of the following chiefs as:

Ezra 2	Ezra 8	Nehemiah 7	Nehemiah 10
9 Bebai (623)	Bebai	Bebai (628)	Azgad
10 Azgad (1222)	Azgad	Azgad (2322)	Bebai
11 Adonikam (666)	Adonikam	Adonikam (667)	Adonijah
12 Bigvai (2056)	Bigvai	Bigvai (2067)	Bigvai
13 Adin (454)		Adin (655)	Adin

The use of Adonijah in Nehemiah 10 rather than Adonikam will be discussed later in this chapter. The numbering sequence reflects a list of twenty chief names in Ezra 2 with only a portion shown for brevity. A basis for difference in descendant counts from Ezra 2 and Nehemiah 7 is not explained, but likely Adonikam (666) vs. Adonikam (667) reflects inclusion of the chief father himself rather than just descendants.

Nehemiah states (Nehemiah 7:5), "…I found a register of the genealogy of them which came up at the first, and found written therein." Since the numbers in Ezra and Nehemiah for other chiefs are not identical, it is evident the records do not reflect an exact copy. They were apparently made at different times for specific purposes, possibly by separate scribes.

An interesting connection in the two numbers associated with the name Adonikam (666) and (667) relates to the significance of numbers six and seven. The number six is associated with man and work as fallen and incomplete. Seven, however, in Hebrew means completion or perfection associated with God and His rest.

As developed following, the name Adonikam reflects someone who rises up. The rise of the Antichrist serves as Satan's final chance to usurp the crown of Jesus Christ who alone conquered death. By raising up a fallen man to rule the world, Satan will conspire to prevent the fulfillment of Christ's kingdom on earth. This beast, introduced in Revelation 13, refers to a head of the beast wounded to death and healed, causing the world to wonder.

The significance of the name does not rest on the name Adonikam itself, but rather the meaning of the name implying to arise, or resurrection. Names and numbers are important based on their meaning tied to other references in Scripture. Specifically, the Lord revealed:

> Here is wisdom. Let him that hath understanding count the number of the beast: for it is the number of a man; and his number is Six hundred threescore and six. (Revelation 13:18)

The association of name and number has no numerological meaning, but rather represents the count or number of persons associated with a prominent individual. Scripture explains Scripture. Unsuccessful attempts at use of numerology have given contradictory outcomes trying to derive 666 from letters of historical or present-day names. That does not mean the Antichrist will be named Adonikam, but the importance rests on the number associated with the name meaning *Lord of Resurrection*. He will rise up from obscurity to world domination, as well as rise up from a mortal wound (Revelation 13:3,12,14).

At the unveiling of the Antichrist, unbelievers will identify with a mark, the name, or the number associated with the name (Revelation 13:17,18) in order to control commerce. The use of tattoos with the number, often combined with images of a pentagram and Satan, are already contemporary symbols of anarchy or rebellion against authority. That is only a foreshadowing of something more sinister to come.

Biblical accounts of Antichrist (Daniel 11:21-23) express how, though vile, he will gain rule by flattery, make a league with many, and work deceitfully to become strong with a small people. In Isaiah (28:14-18), the prophet speaks to Israel about their "covenant with death" and "agreement with hell." This agreement, mentioned by the angel to Daniel (Daniel 9:27), reflects a truce with the nations guaranteeing the security of Israel subject to certain concessions.

The Hebrew word *shabua* for week or seven refers to seventy sevens of years as well as to the final seven years (Daniel 9:27). Other descriptions for the latter half of a final seven label the period as 42 months or

1260 days. The covenant period, patterned after Abraham's covenant with Abimelech at Beersheba (well of the oath, or seven, or completion) forms a framework for the final covenant of seven developing now as the Abraham Accords. The covenant is discussed further near the end of this chapter. This prince to come confirms a covenant with many for seven (Daniel 9:27), though he may not be the originator of the agreement.

The nation Israel insists on a guarantee of security which only God Himself can provide. What allegedly is for their protection becomes Israel's destruction when Antichrist turns, in the middle of the seven-year agreement, to destroy Israel by violating the guarantee of protection (Daniel 11:30-31). The resulting Abomination of Desolation, spoken of by Jesus (Mark 13:14) can be seen by type and shadow through the working of Antiochus-IV Epiphanes (Greek – meaning *resolute protector manifest*) in the books of 1 and 2 Maccabees.

Two books of the Maccabean wars are included among seven Apocryphal books of the Catholic Bible. Though not considered part of the canon of Scripture, they are a valuable historical record of the inter-testament period, which is 400 BCE (Before Christian Era) to 1 CE (Christian Era).

During the Protestant Reformation, Martin Luther translated a German Bible (1534 CE) from the Hebrew and Greek Septuagint, which included apocryphal books. After the death of Martin Luther (1546 CE), Catholic church leadership at the 19th ecumenical Council of Trent adopted apocryphal books from the Latin Vulgate into their Bible as deuterocanonical writings.

Apocryphal books are not part of the Hebrew Scriptures nor the Protestant Bible because of their questionable author, origin and lack of divine inspiration. Events in Maccabees are also detailed in *The Antiquities of the Jews* by Flavius Josephus, a Jewish military leader who became historian to Emperor Vespasian after the destruction of Jerusalem in 70 CE.

Antiochus-IV and The Abomination of Desolation

The Jews, Israel, and the church have variously experienced periods of persecution and tribulation including martyrdom since the first century. Most of the persecution remained localized and of varying intensity.

The Jewish Holocaust during World War II was one of the most severe periods of genocide in recent memory with an estimated death toll of six million European Jews. By comparison, Stalinist era purges and pogroms are estimated to have killed about twenty million persons—for political, ethnic, and religious reasons. Descriptions of the horrors during these atrocities are appalling but few can compare to the documented exploits of Antiochus-IV Epiphanes against the Jews as described in 1 and 2 Maccabees.

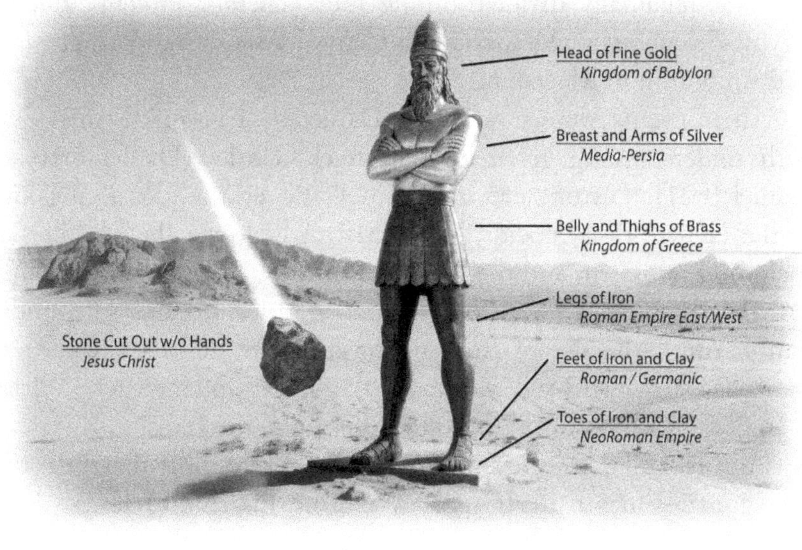

Statue of Nebuchadnezzar's Dream

The Prophet Daniel recorded several dreams and visions regarding a succession of nations ruling Palestine including Babylon, Media-Persia, Greece, and Rome. Identification of these kingdoms aligned with the nations described by the successive metals in the statue of

King Nebuchadnezzar's dream. Each successive kingdom and ruler experienced some aspiration of world domination. The four kingdoms listed were all successful and Daniel's visions from 538 BCE accurately predicted the historical fulfillment.

Of particular interest was the meteoric rise of Greek influence under Philip-II of Macedon. Philip-II, a successful Macedonian military strategist and his wife, Olympias, a devotee to the serpent-handling cult of Dionysos, gave birth to a son, Alexander. His birth coincided with a number of favorable omens.

Olympias, determined to elevate her son Alexander to the throne at all costs, convinced Alexander that he was not fathered by Philip but rather the Greek God Zeus. As a young military commander, Alexander soon demonstrated military prowess by conquering most of the civilized world. But when Alexander the Great (356-323 BC) died suddenly at the age of 32, his world-wide kingdom under Greek rule was regionally divided among four generals; Ptolemy ruled in Egypt, Seleucius in Mesopotamia/Central Asia, Attalid in Anatolia, and Antigonid in Macedon.

These events were revealed to Daniel in a vision (Daniel 7-11) with understanding given by the angel Gabriel (Daniel 8:16 and Daniel 9:21). History provides names and events identifying kings of the Hellenistic Seleucids, kings that reigned over the Middle East, including Jerusalem and the Jewish people.

The dynasty of Seleucid kings ruled Palestine and surrounding nations until the rise of Roman power under Julius Caesar. At the pinnacle of Seleucid rule was Antiochus-IV Epiphanes, who initially tolerated Jewish traditions but grew to despise the Jews.

Antiochus-IV ruled from 175 BCE (Before Christian Era) until his agonizing death in 164 BCE. During his rule of Jewish lands, he decided to consolidate power by embracing Hellenistic Jews who openly adopted Greek traditions.

In order to quell an uprising among faithful Jewish believers, Antiochus-IV outlawed observances and ordered all Jews to worship Zeus. According to the writer of 1 and 2 Maccabees, Antiochus-IV also offended the Jews by instituting a Greek style gymnasium in Jerusalem. In classic Olympian tradition, the Greek word *gymnos* (naked) refers to training and sport conducted completely unclothed. The promotion of Greek games elevated man's achievement and the human body, at which the observant Jews took great insult.

At this provocation, Judah Maccabee led a popular resistance against the Seleucid army. Despite numerous defeats and suffering at the hands of the Greeks, the Maccabees eventually overthrew Antiochus-IV and restored traditional worship to Palestine. Before his death, Antiochus-IV lamented persecuting the Jews (2 Maccabees 9:12).

Of paramount significance during this uprising was the event referred to as the Abomination of Desolation or Holocaust. After King Antiochus-IV decreed the abolition of Jewish customs and worship, he committed an act of conspicuous desecration against the altar and temple to defile it and prevent further use.

The king ordered a swine sacrificed on the holy place of the temple and erected a statue of the Greek God Zeus, having his own image substituted for the face of Zeus. This occurred the month Kislev 25, in 168 BCE. With the temple defiled, Jewish worship ceased until completion of the successful revolt on Yom Kippur 165 BCE.

Sacrifice of a Mother and Her Seven Sons

One account of the horrific persecution under Antiochus-IV regards a mother and her seven sons threatened with death if they did not eat pork (2 Maccabees 7). One by one, as the mother encouraged her sons to disobey, Antiochus ordered the sons to be hacked to death in front of their mother for their refusal. Finally, the mother herself suffered the same death.

Cleansing and restoring the temple took an additional 75 days, commencing the celebration of Hannukah (Festival of Lights) on Kislev 25, 165 BCE. The miracle of the Hannukah lamp holy oil occurred when a day's supply lasted eight days.

Did the followers of Jesus and the multitude of Jews understand the significance of the Abomination of Desolation? Indeed, they were well documented in the writings of 1 and 2 Maccabees as well as Jewish oral tradition. Still, the reference of Jesus to a past event two hundred years prior, as though yet to come, must have seemed a mystery.

Today the Jews know and understand the meaning and significance of the World War II Holocaust. So also, in the time of Jesus the bloody purge of King Antiochus-IV to defile the temple was fresh in their understanding. These trials of the Maccabean revolt, though outrageous in their own right, pale in comparison with coming persecution under the rule of the Antichrist.

Sela (Petra)

During the Olivet discourse, Jesus forewarned believers of a coming Abomination of Desolation. As Jesus spoke,

> When ye therefore shall see the Abomination of Desolation, spoken of by Daniel the prophet, stand in the holy place, (whoso readeth, let him understand:) Then let them which be in Judaea flee into the mountains: ...for then shall be great tribulation, such as was not since the beginning of the world to this time, no, nor ever shall be. And except those days should be shortened, there should no flesh be saved: but for the elect's sake those days shall be shortened. (Matthew 24:15,16,21,22)

In the warning of Jesus, Jews are advised to flee to the mountainous wilderness southeast of Jerusalem upon hearing of the Abomination. Only one hundred forty-four thousand comply and escape in the middle of the seventieth seven. Today there are about 7.7 million Jews in Israel of a total 10.2 million population. Jerusalem's population of about 1 million consists of roughly 600,000 Jews.

Flee where? Jesus and John the Baptist, during their respective ministries, retired beyond the mountain wilderness of Judea southeast of Jerusalem. Past this wilderness the landscape descends rapidly to Jericho and then the Jordan River valley, roughly 1400 feet below sea level.

East of the Jordan (meaning to descend or go down) the topography rises again to a great expanse of mountainous Jordanian desert. There in Jordan lies a region formerly belonging to Moab, Edom, and

Ammon. In the wilderness of Moab, the former city of Petra (or Sela) remains nestled in a secluded valley.

City of Petra in Jordan

The Prophet Isaiah (Isaiah 16:1-5) spoke of this region as Sela, an early name for Petra. Through the prophet Isaiah, God called for outcasts fleeing the spoiler to dwell with Moab until the indignation passes. As Daniel prophesied, Edom, Moab, and Ammon would escape out of Antichrist's control (Daniel 11:41), despite his world domination.

Fifteen hundred years before the ministry of Jesus, the Lord called Joshua (meaning *YHWH* is salvation) to lead Israel over Jordan into the land of promise. In the Hebrew, the names Jesus (*Yeshua*), Joshua, Hosea, and Isaiah all have the same root, origin, and meaning.

Moses, striking the rock a second time to bring forth water (Numbers 20:11), was admonished for failing to speak as instructed. As a consequence, Moses, a figure of the law, died on the other side of Jordan, relinquishing his role to Joshua.

Moses Strikes the Rock the Second Time

The meaning? Law represented in Moses does not lead to the land of promise, only salvation by God's grace represented by Joshua. At the crossing of Jordan (Hebrew *abar* meaning cross over), Joshua instructed the priests to step into Jordan carrying the ark.

Upon entering the waters, Jordan's flow stopped, parted, and rose upon a heap. At the instruction of Joshua, a representative of the twelve tribes selected and carried a stone from Jordan to set up a sign to future generations in Gilgal. This process of descending, parting, and crossing over symbolizes events yet to come at Christ's return in the heavens.

Petra, owing to unique structures carved into the stone walls, has become a modern tourist attraction. Once home to wealthy Nabataeans traders, they developed a thriving community. The Nabataeans sculpted private and public buildings from sandstone cliffs and the city prospered on a lucrative trading route of the Roman Empire.

Gradually Petra came under the control of Rome, giving way to a large Christian church after the rise of Emperor Constantine. In the latter years of the Roman Empire, Petra's influence waned as a place of commerce, culminating with a major earthquake in 363 CE (Christian Era). From then until the 1800s, Petra remained effectively unknown except the small local population of Bedouin inhabitants. Today Petra remains as a popular tourist site for visitors from Israel and Jordon.

The Lord also revealed (Revelation 12) imagery of the woman (a type of Israel) bringing forth a man child (a type of Christ). The woman escapes the serpent's destruction, fleeing to the wilderness (Revelation 12:12-17) in the middle of the seven years for a time, times, and half a time (3-1/2 years, 42 months, 1260 days).

Through this time of Great Tribulation, pervasive in the earth, one-hundred-forty-four thousand as yet unredeemed Jews steal away out of Israel to the wilderness, hidden from the Antichrist. Isaiah (Isaiah 10:21-23) and the Apostle Paul (Romans 11:5) raise the specter of a small remnant of Jews who escape, hide, and are redeemed at the appearing of Jesus Christ in person.

This event fulfills the prophetic promise of preparing, separating, and then sealing of one-hundred-forty-four thousand from twelve tribes prophesied in Revelation 7 prior to the opening of the seventh seal and the beginning the Day of the Lord's Wrath.

The redemption of this remnant at the close of seven years (7 x 360 = 2520 days) brings to completion a literal fulfillment of the Day of Atonement (Yom Kippur). For the church of true believers, a unique day approaches within the latter 3-1/2-year window of the seventieth seven or *shabua*. That unknown day and hour, comes suddenly, as the announcement of a new moon on the festival Rosh Hashana, proclaimed by the sound of a trumpet and signal fires, upon sighting of the new crescent moon at sunset.

Michael the Restrainer

Canon of Scripture records three chief angels or archangels by name; Michael (meaning who is as God), Gabriel (meaning warrior of God), and Heylel (meaning shining one or Lucifer). Both Gabriel (Daniel 9:21, Luke 1:26) and Michael (Daniel 12:1, Revelation 12:7) receive multiple references and appear in Old and New Testament accounts.

However, the Hebrew language name Heylel has only one Old Testament mention (Isaiah 14:12). Most Christians are unaware of the name Heylel because of the use of Lucifer from the Latin Septuagint translation. Notably, of the three archangels, Heylel's name is the only one of the archangels whose name references himself rather than God.

Jerome, one of the early church fathers, substituted the Latin name Lucifer (morning star, shining one) rather than the Hebrew name. Other names of angels based on non-canonical literature such as the Book of Enoch are of uncertain origin and not addressed here. As noted in Chapter 12, the Arabic name Zohran (morning star, to shine) has the equivalent meaning of Hebrew Heylel.

Descriptions for this third archangel address him as leader of worship; a proud, pompous, and self-willed figure. In Isaiah 14, Heylel boasts "I will" five times:

> For thou hast said in thine heart, I will ascend into heaven, I will exalt my throne above the stars of God: I will sit also upon the mount of the congregation, in the sides of the north: I will ascend above the heights of the clouds; I will be like the most High. Yet thou shalt be brought down to hell, to the sides of the pit. (Isaiah 14:13-15)

Isaiah 14 and Ezekiel 28 reference the fall of this third archangel employing titles King of Babylon and Prince of Tyre respectively. These two chapters lay out his fall from grace for pride and leading a third of the angels in rebellion against God. Most portrayals of this third archangel cast him in his fallen role as Satan (meaning adversary or accuser), Devil (meaning slanderer), and dragon. All that see the fallen archangel Heylel at his appearing will be astonished saying,

> They that see thee shall narrowly look upon thee, and consider thee, saying, Is this the man that made the earth to tremble, that did shake kingdoms; That made the world as a wilderness, and destroyed the cities thereof; that opened not the house of his prisoners? (Isaiah 14:16, 17)

In Daniel 10, the prophet speaks of three weeks abstention from pleasant bread, flesh, and wine while he sought the meaning of a vision. During this time, a vision appeared to Daniel saying:

> Now I am come to make thee understand what shall befall thy people in the latter days: for yet the vision is for many days. (Daniel 10:14)

And:

> But I will shew thee that which is noted in the Scripture of truth: and there is none that holdeth with me in these things, but Michael your prince. (Daniel 10:21)

Here, the Hebrew word *chazaq* (meaning to strengthen, hold, or restrain) reflects Michael's role as restrainer of evil against the people Israel. Daniel 11 develops the prophetic foreshadowing of Antiochus-IV Epiphanes and his campaign against the Jews during the period of Maccabean rule.

As the narrative transitions from Greek rule (about 168 BCE) to the latter days in Daniel 11:36, Michael enters the picture again as protector of Israel. However, at this time Michael takes a different stance from his normal role as defender:

> And at that time shall Michael stand up, the great prince which standeth for the children of thy people: and there shall be a time of trouble, such as never was since there was a nation even to that same time: and at that time thy people shall be delivered, every one that shall be found written in the book. (Daniel 12:1)

In this verse, the Hebrew word for stand up *amad* (meaning stand or stand still) implies Michael stands aside from his role as restrainer and permits the enemies of the Jews to attempt the destruction of Israel. God Himself through the action of Michael guarantees the protection of the nation and the people of Israel (Daniel 10:21).

As a consequence of this covenant with death, Israel will sign an agreement under duress with her arch-enemy, guaranteeing the protection only God can give. Upon Israel's apostacy, God gives them, "what you demanded" (or Hebrew *shaul*). The Antichrist then violates the agreement by turning to destroy the nation and people.

In the final days of King Saul's rule, he sought unsuccessfully to find and destroy David, the anointed heir apparent to the throne. Though given two opportunities to destroy King Saul, David would not raise his hand against the king. Instead, David waited for God Himself to reciprocate and judge Saul in the proper time.

Speaking through Nathan the prophet (2 Samuel 7:11-17, 1 Chronicles 17:7-15), God denied David's plan to build God a house, promising instead to build David a house founded upon a yet to be born son.

That promise saw fulfillment first in David's son Solomon, and finally in Jesus, Son of David. One purpose of the Day of the Lord's Wrath remains to bring the remnant to faith in the true Messiah of the Jewish people.

The final day of the seven-year *shabua* on Yom Kippur brings in the remnant of the Jews and closes the Kingdom of God. Satan, the embodiment of the fallen angel Heylel, wants nothing more than to delay this inevitable outcome. Repentance and salvation of the remnant of the Jews signals the end of Satan's attempt to usurp the crown and rule over earth in order to prevent Christ's Kingdom.

In His parables Jesus used types, shadows, and metaphors to create a mystery. Whether in Scripture or a novel, mysteries present a challenge to be unraveled yet remain hidden to the careless reader. As adversary, Satan performs in the role of Jester. By playing the part of the cunning fool, he interjects himself into the plot to distract the unwary.

The Beast

History and Scripture are replete with foreshadowing of the final antagonist character, Antichrist. This "Little Horn" (Daniel 7:8) comes out of obscurity and quickly usurps authority belonging to others. A few of these historical figures in their time were thought by the church to be the Antichrist. In retrospect, these were only a prefigure of the person called the "Little Horn".

1. Antiochus-IV Epiphanes, (216-164 BCE)
2. Emperor Nero, (37–68 CE)
3. Emperor Diocletian, (244-316 CE)
4. Emperor Napoleon Bonaparte, (1769-1840 CE)

5. Adolf Hitler, (1889-1945 CE)

One characteristic description of the Beast is;

> ...one of his heads as it were wounded to death; and his deadly wound was healed: and all the world wondered after the beast. (Revelation 13:3).

As Jesus Christ was raised from the dead, so the imagery of the name Adonikam reflects rising up. It shows he will rise from a deadly wound according to the meaning of his name "Lord of Resurrection" or "The Lord Rises Up." And what of the use of Adonijah (Nehemiah 10:16) instead of Adonikam? Occasionally alternative names are used or assigned as in the case of Pharaoh Necho giving Eliakim the name Jehoiakim (2 Kings 23:34).

Here is an interesting connection. Prior to the promotion of Solomon to the throne, Adonijah (meaning my lord is Jehovah) sought to elevate himself and usurp the kingdom (1 Kings 1:8). Though pardoned by Solomon for a time, a subsequent request by Adonijah to marry David's nurse Abishag was interpreted as seeking the crown, resulting in Adonijah's execution for treason.

This scenario with names and imagery of Adonikam and Adonijah provides a foreshadowing of the rebellion of Satan and the Antichrist. He must experience a deadly wound and be healed. Then Satan, through his resurrected minion the Antichrist, will attempt to usurp Christ's rightful crown and dominion in the final days before the catching away, and eventually be thwarted by the wrath of God.

In the coming peaceable reign of Christ, when he returns to rule for a thousand years, Satan must be bound and the Antichrist (or beast) cast into the lake of fire with the false prophet (Revelation 19:20). At the same time Christ slays the army following the Antichrist with the sword (Word of God) out of his mouth (Revelation 19:21).

However, after the thousand-year reign, Satan must be loosed for a final rebellion among unbelieving nations. Those unregenerate have their lives prolonged during the millennial reign of Christ as declared in Daniel 7:

> I considered the horns, and, behold, there came up among them

> another little horn, before whom there were three of the first horns plucked up by the roots: and, behold, in this horn were eyes like the eyes of man, and a mouth speaking great things. I beheld till the thrones were cast down, and the Ancient of days did sit, whose garment was white as snow, and the hair of his head like the pure wool: his throne was like the fiery flame, and his wheels as burning fire. A fiery stream issued and came forth from before him: thousand thousands ministered unto him, and ten thousand times ten thousand stood before him: the judgment was set, and the books were opened. I beheld then because of the voice of the great words which the horn spake: I beheld even till the beast was slain, and his body destroyed, and given to the burning flame. As concerning the rest of the beasts, they had their dominion taken away: yet their lives were prolonged for a season and time. I saw in the night visions, and, behold, one like the Son of man came with the clouds of heaven, and came to the Ancient of days, and they brought him near before him. And there was given him dominion, and glory, and a kingdom, that all people, nations, and languages, should serve him: his dominion is an everlasting dominion, which shall not pass away, and his kingdom that which shall not be destroyed. (Daniel 7:8-14)

God demonstrates His righteousness by prolonging the life of the rebellious who remain alive after the Day of the Lord's Wrath. Scripture portrays the coming thousand-year reign of Christ with believers as a strict though compassionate dominion.

Together, Christ and the church comprised of Old and New Testament believers rule over the unregenerate nations of the world as demarcated by the stern Kingdom of David and peaceable Kingdom of Solomon.

> Thou shalt break them with a rod of iron; thou shalt dash them in pieces like a potter's vessel. Be wise now therefore, O ye kings: be instructed, ye judges of the earth. Serve the Lord with fear, and rejoice with trembling. Kiss the Son, lest he be angry, and

> ye perish from the way, when his wrath is kindled but a little. Blessed are all they that put their trust in him. (Palms 2:9-12)
>
> Let the saints be joyful in glory: let them sing aloud upon their beds. Let the high praises of God be in their mouth, and a two-edged sword in their hand; to execute vengeance upon the heathen, and punishments upon the people; to bind their kings with chains, and their nobles with fetters of iron; to execute upon them the judgment written: this honour have all his saints. Praise ye the Lord. (Psalms 149:5-9)

After the millennium, these nations, though living under benevolent rule and the reign of Christ and the saints for a thousand years, when given the alternative of submitting to Christ or Satan, choose the latter. But then, just as David slew the Amalekite (2 Samuel 1:15,16) and Solomon destroyed Adonijah (1 Kings 2:24,25), Satan must then be finally vanquished. Satan will then be cast into the lake of fire, which is the final death closing the millennium (Revelation 20:10).

Yet all, righteous and rebellious, must ultimately stand before the Great White Throne judgment of God in heaven (Revelation 20:11-15). There the books will be opened and the saints justified by Christ's sacrifice while the words and works of the ungodly bring their judgment.

Cosmic Signs

Returning to the opening of the seals, a parallel can be drawn between the description of cosmic signs in the Olivet Discourse of Matthew 24 and the opening of the sixth seal in Revelation 6:12. Jesus stated regarding the period of Great Tribulation commencing after the Abomination of Desolation, that no flesh would be saved except those days should be shortened (Matthew 24:22). Referring to this abbreviation of the Great Tribulation, Jesus declared,

> Immediately after the tribulation of those days shall the sun be darkened, and the moon shall not give her light, and the stars shall fall from heaven, and the powers of the heavens shall be shaken: And then shall appear the sign of the Son of man in heaven: and then shall all the tribes of the earth mourn, and they shall see the Son of man coming in the clouds of heaven with power and great glory. And he shall send his angels with a great sound of a trumpet, and they shall gather together his elect from the four winds, from one end of heaven to the other. (Matthew 24:29-31)

The parallel passage at the opening of the sixth seal paints a similar picture of a cosmic disturbance:

> And I beheld when he had opened the sixth seal, and, lo, there was a great earthquake; and the sun became black as sackcloth of hair, and the moon became as blood; and the stars of heaven fell unto the earth, even as a fig tree casteth her untimely figs, when she is shaken of a mighty wind. And the heaven departed as a scroll when it is rolled together; and every mountain and island were moved out of their places. (Revelation 6:12-14)

These two passages summarize the primary plot of this book. The blessed hope (Titus 2:11-14) refers to a joyous expectation of the Lord Jesus Christ appearing. Christ will indeed come again as He departed. First, He will appear in the heavens to raise the righteous dead and catch away believers from the present fallen world. Then the wrath of God will fall during a period called the Day of the Lord.

This is a great mystery. Yet there are aspects of His coming known because they are discernable. Scripture provides a revealed framework of cosmic events surrounding Christ's return. These are not isolated occurrences, but rather part of a sequence of signs emanating from the initiation of a single action by God Himself. These events are not natural in the sense of resulting from a physical cause.

Rend the Heavens ...and Come Down

Isaiah 64 opens with a plea that ties into the description of heaven and earth at the creation.

> Oh that thou wouldest rend the heavens, that thou wouldest come down, that the mountains might flow down at thy presence. (Isaiah 64:1)

In the same way, David prophesied in Psalm 18:

> Then the earth shook and trembled; the foundations also of the hills moved and were shaken, because he was wroth. There went up a smoke out of his nostrils, and fire out of his mouth devoured: coals were kindled by it. He bowed the heavens also, and came down: and darkness was under his feet. ...Then the channels of waters were seen, and the foundations of the worlds were discovered at thy rebuke, O Lord, at the blast of the breath of thy nostrils. He sent from above, he took me, he drew me out of many waters. He delivered me from my strong enemy, and from them which hated me: for they were too strong for me. (Psalm 18:7-9, 15-17)

And also in Psalm 144:

> Bow thy heavens, O Lord, and come down: touch the mountains, and they shall smoke. (Psalm 144:5)

This opens the door to the Day of the Lord. In addition to the earlier reference in Matthew 24, cosmic events related to a darkened sun and moon are prophesied elsewhere in Joel, Zephaniah, 1 and 2 Thessalonians, and Revelation.

Various descriptions including signs in the sun, moon, stars, and earth seem somewhat disconnected on the surface. Yet upon closer examination, they all emanate from the heavens rending open to introduce the return of Christ.

Going back to the earlier discussion, the created universe can be described as a spacetime fabric, with mass and energy embedded and interacting as represented by Einstein's relativistic field tensor. This fabric corresponds to the description provided in Matthew 24. Isaiah and Jeremiah denote the heavens rending, stretched, and rolled up.

Following is a sequence of observable happenings resulting from the rending open of the heavens.

The Earth Trembles

In recent years a scientific endeavor called LIGO-Virgo collaboration has been developed to detect major gravitational disturbances in the universe. LIGO (US based pair of Laser Inferometer Gravitational Wave Observatories) and Virgo (complementary European observatory) sense gravity waves originating from beyond the earth.

Identifying subtle gravitational shifts emanating from cosmic disturbances requires a unique sensing methodology referred to as Laser interferometry with lasers targeted at sensors miles apart for sensitivity.

By sensing these disturbances at two widely separated earth-based locations, LIGO-Virgo can determine whether the event originates from beyond the earth and from what location. A celestial disturbance can then be identified with an observed astronomical event.

References to the beginning of the Day of the Lord include Joel (3:15-16), Isaiah (13:10-13), Matthew (24:29), Mark (13:25), Luke (21:26), Revelation (6:12), and Hebrews (12:26). Each Scripture describes a great shaking of heaven and earth. Many of these Scriptures denote tearing or rolling up of the heavens like a scroll following a terrestrial great earthquake. The implication of rending the heavens precipitates a worldwide shaking on earth. As the Lord said:

> ...Yet once, it is a little while, and I will shake the heavens, and the earth, and the sea, and the dry land; and I will shake all

> nations, and the desire of all nations shall come: and I will fill this house with glory, saith the Lord of hosts. (Haggai 2:6,7)

At the first indication of a major earthquake, virtually everyone makes their way outside because of the risk of a building collapse. At the opening of the sixth seal before the Day of the Lord, a worldwide major quake will have the result of forcing populations to immediately go outside. Visible to all the earth there will be a rapidly unfolding succession of unprecedented cosmic events.

Before looking at these events individually, consider how this will affect everyone in the world simultaneously. While the rending of the heavens and subsequent quaking will touch everyone around the world immediately, can the same be said for visually observable cosmic events?

First, most of our understanding of the heavens is based on observable measurement rather than physical sensing. This includes visible electromagnetic radiation (light spectrum from infrared to ultraviolet) as well as radio waves (longer wavelength) and x-rays or gamma rays (short wavelength). Until recently the heavens were perceived only by visible radiation. Understanding these observables utilizes mathematics (symbolic logic) and theoretical modeling to explain a phenomenon or to predict a future event under specified conditions.

Only within the past sixty years, space probes have travelled to the moon, Venus, Mars, asteroids, as well as interstellar space to make measurements that do not solely rely on electromagnetic waves.

Results are still returned in the form of electromagnetic signals. In the case of lunar landings, some physical samples have been returned for inspection. When the Day of the Lord unfolds, will the world depend on television, video camera, or cell phone transmissions to observe these events?

To answer that question, consider what occurred in the battle supporting the Gibeonites against the Amorite nations (Joshua 10:12-13). Joshua spoke to the Lord and commanded the sun and moon to stand still. Verse 14 says, "And there was no day like that before it or after it, that the Lord hearkened unto the voice of a man."

The names Jesus, Isaiah, Joshua, and Hosea all derive as an abbreviation from the Hebrew name Jehoshua (meaning Jehovah saved).

Joshua declared, "...Sun, stand thou still upon Gibeon; and thou, Moon, in the valley of Ajalon." (Joshua 10:13)

God Himself responded to the assertion of Joshua as a type of Christ to foreshadow events at the opening of the sixth seal. Similarly, in Hezekiah's interaction with Isaiah, he was given the option for a sign of the sun's shadow lowering or returning ten degrees on the sun dial of Ahab (Isaiah 38:8).

Many biblical narratives describe miraculous events. As written, the battle in Joshua continued while the sun and moon appeared to remain fixed in the sky for about a day. What was happening in the heavens?

For anyone with a knowledge of orbital mechanics, that is problematic. While it can be argued, God is not limited by physical processes, and can do anything He wants, we know stopping the movement of the sun and moon would involve quite a complex set of maneuvers.

Technically the earth would need to decelerate and then reaccelerate in this case. With the earth's circumference of roughly 25,000 miles rotating at one revolution each 24 hours, that is a surface velocity of just over 1000 miles per hour. The resulting tidal shift of a rotational stop/start would be traumatic—let alone all the other implications of such an event. Something else must be going on.

In the story of Hezekiah and Isaiah, the offered sign was framed as the shadow returning, rather than the earth turning back or sun changing directions. If we consider spacetime as a fabric, the sun and moon can appear to stand still simply by a distortion of the spacetime fabric.

That is no less a miraculous maneuver, but makes much more consistent sense in the description of what happens in the books of Joshua and Isaiah and at the end of the age. In each case, spacetime distorts by stretching or rending of the heavens.

As the heavens are rolled up as a scroll, light and electromagnetic radiation in transit will be distorted and appear as if the sun, moon and stars are moving across the heavens. As described in Matthew 24:29, "...the stars shall fall from heaven...."

The movement of light from the stars will appear as though they are falling across the skies. Coincidently, recent observation of distant galaxies with the James Webb Space Telescope (JWST) illustrates how light bends around large masses, in a process termed gravitational lensing, a distinct but similar occurrence.

Consider a further characteristic of mass, spacetime, and the movement of light or radiation based on Einstein's relativistic field equations. Objects are characterized by mass, and an associated contribution to spacetime. The influence of all other mass in the universe also acts on every object by contributing to the vacuum energy density of space characterized by spacetime fabric or tension on the continuum.

Even without the influence of other mass, each object initiates its own spacetime contribution. This characteristic results in relativistic differences of spacetime measure when an object moves at or near the speed of light. At one point in his development of the tensor field equation, Einstein added a Cosmological Constant to create a relationship which predicted a non-expanding universe based on the equation. It was not until Edwin Hubble demonstrated the universe was expanding based on the redshift of light, that Einstein declared his use of the constant a mistake.

If the bound of the heavens and fabric of space is rent, and rolled up, that is a non-local influence. Local spacetime created by sun, earth, and moon still provide dominant influence on localized spatial fabric.

But light in transition from stars and sun particularly are shifted. And as the spatial fabric of the heavens quickly rolls back, light from the stars, sun, and moon appear to move across the sky with everyone across the world witness to the result. Look at more implications of the heavens rolling away.

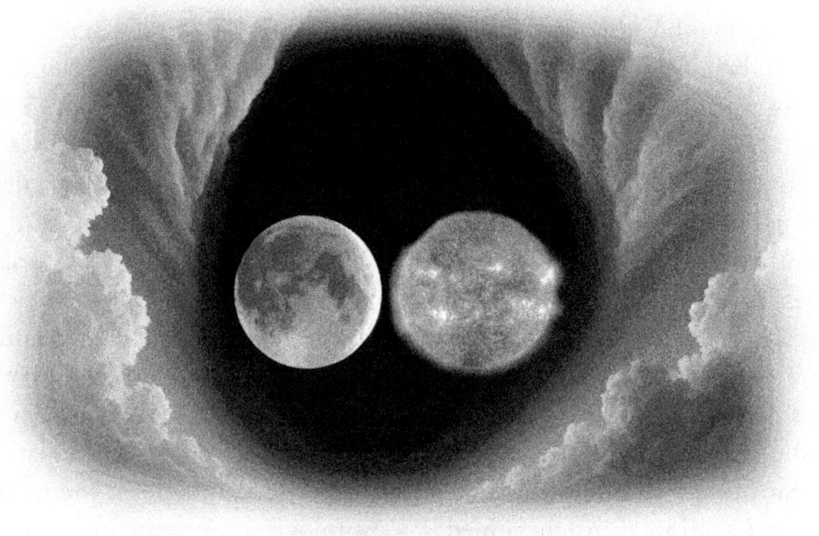

The Heavens Roll Away

Arch of the Heavens

In the creation account there are several descriptions about the nature of the heavens. At the beginning, heaven and earth are described as without form, void, and dark while the Spirit of God moved on the face of the waters. With the Word of God came the creation of light and the separation of light and darkness defining the first day and night. Again, the Word of God spoke into existence a firmament within waters on the second day, dividing the waters above and below.

> And God said, Let there be a firmament in the midst of the waters, and let it divide the waters from the waters. And God made the firmament, and divided the waters which were under the firmament from the waters which were above the firmament: and it was so. And God called the firmament Heaven. And the evening and the morning were the second day. (Genesis 1:6-8)

What is meant by waters? The Hebrew term *waters* or *mayim* (meaning water or fluid) appears straightforward. The Hebrew for *heaven* or *shamayim* means "sky" or "lofty waters." On the fourth day God said, "…Let there be lights in the firmament of the heaven to divide the day from the night; and let them be for signs, and for seasons, and for days, and years." (Genesis 1:14)

The sun, moon, and stars were set (from the Hebrew term *nathan* meaning given) to establish light on the earth and to create the division of day and night. Light was not created in process, so the firmament would have stretched out on the fourth day as the stars formed. The expression, "…a firmament in the midst of the waters, and let it divide…" paints an image of the extent of the heavens bounded at its limit and then stretched out.

When God stretched out the heavens, the composition bounding the perimeter of the sphere was not described beyond waters. The use of the Hebrew word *shamayim* or "lofty waters" does not qualify an exact nature or understanding of the waters constituting the outer limit of the firmament. In my previous book *Rend the Heavens*, I discussed the discovery and description of the Cosmic Microwave Background (CMB) radiation as evidence of a bound to the universe.

The best determination of this perimeter would be something like a boundary of water, or form of matter providing considerable mass as a boundary of the heavens. Regardless of the boundary makeup, the specifics are secondary to the central point that it serves as the mechanism of mass with means to stretch out the cosmos.

Because Scripture does not provide a clarification of the makeup, the term "waters" for this bound suffices. Other Scriptures support the Genesis 1 account of waters bounding the cosmos. Elihu reproved Job with the question:

> Hast thou with him spread out the sky, which is strong, and as a molten looking glass? (Job 37:18)

In the Psalms, the waters above the heavens are referenced:

> Praise him, ye heavens of heavens, and ye waters that be above the heavens. (Psalm 148:4)

Also, in Revelation:

> And before the throne there was a sea of glass like unto crystal... (Revelation 4:6)
>
> And I saw as it were a sea of glass mingled with fire... (Revelation 15:2)

And in 2 Peter:

> And saying, Where is the promise of his coming? For since the fathers fell asleep, all things continue as they were from the beginning of the creation. For this they willingly are ignorant of, that by the word of God the heavens were of old, and the earth standing out of the water and in the water: Whereby the world that then was, being overflowed with water, perished: But the heavens and the earth which are now, by the same word are kept in store, reserved unto fire against the day of judgment and perdition of ungodly men. (2 Peter 3:4-7)

Beyond this bound there would be nothing of the physical world; no space, no time, no mass, no light, at least not in a natural sense. But these "waters" create gravitational tension on the interior heavens. A key visual observation is the redshift of light which increases with distance. To express this in a more technically correct manner, mass causes a distortion of spacetime, which in turn results in light and mass moving in a way that appears to give rise to gravitational force.

Cosmologists currently hypothesize this results from dark matter and dark energy distributed within the universe. The intent is to explain observed acceleration toward the peripheral bound or edge of the cosmos. The creation account presents these bounding waters as the matter continuing to stretch out the fabric of spacetime and resulting redshift of light.

Cosmologists do not have clear and convincing evidence of the existence of any other form of dark matter or dark energy; only that something provides a mechanism like Einstein's cosmological constant to explain observed behavior of the universe. The Lord took Job to task when he asserted that God owed him an explanation, saying, "Where wast thou when I laid the foundations of the earth?" (Job 38:4).

Scripture presents a compelling picture of the unveiling of creation, the stretching out of the heavens, and an explanation for the observed universe. In addition, this serves as a means to illustrate how the events of the Lord's return unfold consistent with the biblical account. Mass constituting the outer bound to the cosmos explains the observed stretching rather than the need to postulate dark matter. Also, the boundary emits radiation at the average color temperature characterizing the CMB, close to absolute zero.

Sun as Sackcloth

God's purpose at the opening of the sixth seal represents a work to initiate the removal of the faithful and dead in Christ as well as begin judgment during the Day of the Lord's Wrath. It does not constitute a deterministic outcome of events already started at the bounding surface, but of an immediate response to God's fiat action, rolling out the cosmic scene to everyone across earth in rapid succession.

As space-time stretches with rending of the heavens, sunlight shifts to infrared. Also, the sky becomes black due to a darkened sackcloth sun and the moon turns blood red. At the same time stars appear to traverse the sky and the scene rolls quickly around the earth. How does this materialize from the stretching of the heavens?

First consider that each celestial object has a different spectral signature. Light is emitted based on temperature and composition or it is reflected subject to light source, composition, and reflectivity. There are other factors such as object size, redshift due to motion, or the effect of clouds, for example, on Jupiter and Saturn. In addition, light experiences absorption lines resulting from elements on the surface or in the path to earth.

The nature of these absorption lines gives information about the surface composition or material in interstellar space. Our sun has its own unique spectral signature measured in space above earth as well as on the earth's surface. As with all stars, the sun's spectrum ranges in wavelength and intensity from infrared to ultraviolet, with the preponderance of solar energy in the visible to ultraviolet.

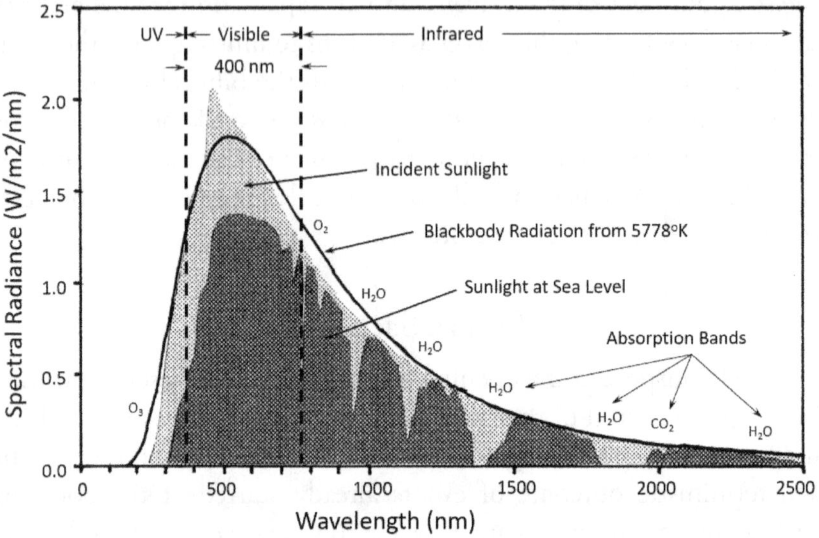

Chart of Solar Radiation Spectrum

This occurs as a function of temperature and makeup of the sun's surface and corona where radiation originates. At the surface of earth, specific wavelengths of light exhibit lower intensity resulting from energy absorption and scattering.

Ultraviolet light dispersion due to oxygen (O_2) and ozone (O_3) represents the most obvious example of scattering, giving rise to blue skies in daylight. Outside the atmosphere, solar radiation contains damaging levels of UV radiation, and lacking scattering, the sky appears totally black rather than blue.

Moving to lower altitudes skies transition from black, to indigo, to blue as more oxygen molecules scatter UV light. Atmospheric conditions at the Dead Sea (~400 m below sea level) create unique circumstances of low humidity and low UV light favorable for skin phototherapy. When frequencies of light shift due to the stretching of spacetime, unique changes take place as the preponderance of ultraviolet transitions toward infrared.

Loss of blue wavelengths means the sky in daylight becomes completely black, typical of what is experienced outside the atmosphere from the stratosphere and beyond. Also, as sunlight becomes mostly infrared, the sun takes on a dark red hue in a black sky. So, a reddish-black sun transitions across a dark sky as the heavens roll quickly away around the world. That reflects the biblical description in both the Old Testament and New Testament accounts of characteristic signs preceding the Day of the Lord. Next, what happens to the appearance of the moon?

Blood Red Moon

Recent prophetic teachers have spoken about a series of so-called "blood moons" prior to the Lord's return. A blood moon in the popular sense refers to a lunar eclipse or other naturally occurring events. These notions do not match biblical narratives prior to the Day of the Lord where a blood red moon appears in conjunction with other celestial events such as the sun darkened like sackcloth. Descriptions of a blood red moon in Isaiah, Joel, Matthew, and Revelation represent something entirely unique. First, consider what happens in the context of events at the Lord's return in the heavens just prior to the Day of the Lord's wrath. The circular chart of the yearly Jewish calendar pictured here again highlights the Feasts of the Lord.

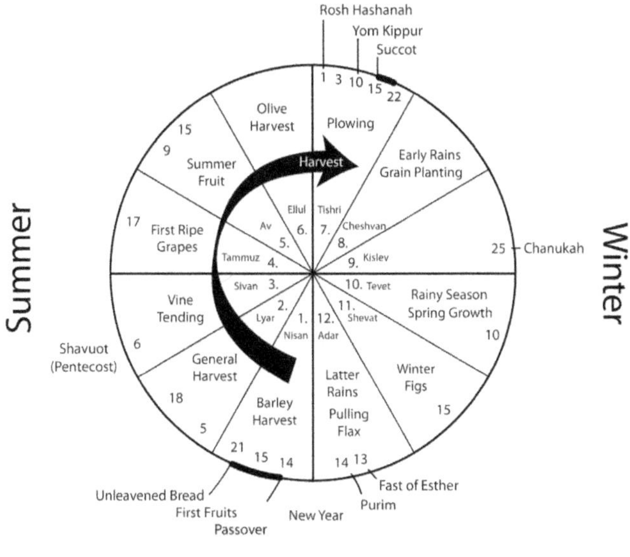

Circular Chart of Jewish Yearly Calendar

Feasts of Passover Season

Of seven "Festivals of the Lord" in Leviticus 23, four have been completed or fulfilled. Each of the seven is a symbolic representation based on the past as a precursor of the future. Passover Season, instituted symbolically in the exodus from Egypt had fulfillment with Christ's sacrifice on the cross. At the celebration of Passover, three festivals are satisfied. Passover, First Fruits, and Unleavened Bread had their symbols in Egypt and their completion in Christ during Passover season at Christ's Passion and resurrection. These three in addition to the Festival of Weeks were fulfilled in the year 33 CE as explained earlier.

Festival of Weeks (Shavuot)

In Exodus 19, on the third month and third day, the people prepared themselves to receive the law by the hand of Moses. Coming of the law written in stone fifty days after Passover was symbolic of the work of the Holy Spirit when, on the Day of Pentecost (Weeks), the law was revealed to the heart. Hence the fourth festival also had fulfillment symbolically and literally at the Jewish Festival of Weeks in 33 CE.

Feast of Trumpets (Rosh Hashanah)

For this reason, fulfillment of Christ's return in the heavens would be expected at the Festival of Trumpets or Rosh Hashanah which means head of the year. This begins the first day of the seventh month of the Jewish religious calendar (first day of the first month of the civil year). How does this reconcile with what Jesus declared?

> But of that day and hour knoweth no man, no not the angels of heaven, but my Father only. (Matthew 24:36)

Consider that the two days of Rosh Hashanah, though beginning a set day each year based on the current Jewish calendar (163 days after Passover), was not traditionally identified until the High Priest certified the first new moon of the civil year. Rosh Hashanah begins the evening of the first day and ends the evening of the third day (equaling two days).

Uncertainty of the identification of a Rosh Hashanah new moon (first sliver) resulted in two days observance. Observing the moon prior to appearance of the new moon becomes difficult near the sun. Apart from new moon occurrences associated with a partial or full solar eclipse, a new moon would appear as a sliver of crescent after sunset.

The moon's orbit of earth lies in a plane tilted about five degrees from the ecliptic, which is Earth's plane of rotation about the sun. Therefore, the moon often appears out of plane from the sun when observing a new moon crescent. Even under ideal conditions just after sunset, observing the new crescent moon remains quite difficult to see.

Therefore, no one "knew the hour or day" until signaled by the High Priest based on an observable sign seen by two witnesses. Though the "times and seasons" (I Thessalonians 5:1) were known then, as they are observed today, the new moon depended on the High Priest announcement, "*Mekudash! Mekudash!*" meaning "It is sanctified!".

Once declared near sunset, bonfires on hilltops and the blowing of trumpets signaled Rosh Hashanah to the nation of ancient Israel. A seeming inconsistency of the new moon at Rosh Hashanah and these cosmic events would be the description of a blackened sun and blood red moon with rending and rolling up of the cosmic spacetime fabric.

Normally a new moon cannot be seen with ease against the backdrop of the sun, unless there is reflected earthlight to illuminate the moon. This difficulty is compounded by the brilliance of the sun, but not if the sun's light is darkened by the frequency or wavelength shift to infrared.

Now the challenge is still one of reduced sunlight intensity or luminosity because the light spectrum has shifted from ultraviolet to infrared. Though the moon and sun are roughly in conjunction, the dilemma is not resolved by the conditions of a solar eclipse.

It would seem difficult for the moon to appear so conspicuous without further explanation. Is it possible the moon has greater reflectance to infrared wavelengths than visible or ultraviolet? More to come on this shortly.

The first clear and consistent appearance of believers caught away to heaven occurs after sealing of the one-hundred-forty-four thousand who are protected against wrath to come. John declared the assembly he saw in heaven:

> After this I beheld, and, lo, a great multitude, which no man could number, of all nations, and kindreds, and people, and tongues, stood before the throne, and before the Lamb, clothed with white robes, and palms in their hands; and cried with a loud voice, saying, Salvation to our God which sitteth upon the throne, and unto the Lamb. (Revelation 7:9-10)

All classes or categories of people are included with two notable exceptions. Absent are mention of "all beliefs or faiths," nor provision for "all moral distinctives," i.e. no thieves, no murderers, no immoral, no blasphemers. "Beliefs" and "actions" are two qualities of character that separate sheep and goats; those who inherit eternal life verses those who inherit death based on matters of choice. The Hebrew *Sheol* (Hell) implies, "That's what you demanded."

Feast of Yom Kippur

The Day of Atonement (Yom Kippur) follows Rosh Hashanah on the tenth day of the seventh month, representing the highest holy day of the Jewish calendar. A unique feature of Yom Kippur relates to the day of general repentance, sacrifice, and atonement of the Jewish people. It represents a symbolic yearly celebration and a final day of the seven-year period. On that day God brings the remnant of one-hundred-forty-four thousand face to face and the knowledge of Jesus Christ as Jewish Messiah.

Basis for the Jewish festival calendar rests upon a "luni-solar" system. A lunar month based on astronomical observation is approximately 29-1/2 days, yielding a 354-day lunar year. In order to match the solar year of roughly 365-1/4 days, a leap month Adar is added seven times every nineteen years. In the book of Daniel, as well as elsewhere, an even 30-day month is used for a 360-day prophetic year.

In Daniel 9, God revealed through the archangel Gabriel, a period of seventy-sevens (490 years) or seventy *shabua* (meaning seven) are determined upon Jerusalem (Daniel 9:25-27). Of these, seven plus sixty-two weeks of years (or 69 x 7 = 483 years) from the command to restore Jerusalem until revelation of Messiah the Prince. Here a year entails the 360-day Jewish prophetic calendar year.

Prophetically, the Jewish calendar period matches to the day from Artaxerxes decree to Nehemiah for rebuilding Jerusalem to the triumphal entry of Jesus on Palm Sunday, 33 CE (Christian Era, Zechariah 9:9). That day represented completion of the promise for the revelation of the Messiah.

There remains a final week or *shabua* of years upon God's people the Jews. Regrettably, the final *shabua* frequently has been referred to as "The Tribulation." For thousands of years God's people, both Jewish and Christian, have experienced tribulation.

At any given historical time and world region, believers suffer persecution representing loss of livelihood, family, finances, homes, places of worship, health, and life itself. Though statistics vary depending on the definition of persecution, a large percentage of believers today face what can be called true tribulation, though limited in space and time.

Recent reports of Christian beheadings represent a foreshadowing of things to come at the opening of the fifth seal in Revelation 6:9-11. These three verses encapsulate the period Jesus referred to as a time of "Great Tribulation" (Matthew 24:21,22), when both Christians and Jews throughout the world will suffer affliction, torture, and death of a greater scale than the Maccabean or Nazi Holocausts.

That is nothing new. Now, in the twenty-first century, Christians and Jews have been murdered and beheaded in Syria, Nigeria, Israel, India, and many other nations. As related in the book *Rend the Heavens*, I have faced several near-death experiences in travels around the world.

Great tribulation commences with the introduction of the Abomination of Desolation defiling sacrificial service in a temple or tabernacle, and interrupted at the sixth seal with the appearance of Jesus Christ in the heavens. With opening of the seventh seal, action shifts dramatically from the Antichrist inspired great tribulation to *YHWH* executed great wrath upon the ungodly.

This restricted period of months or "little season" (Revelation 6:9-11) encapsulated in these three verses starting the mid-point of the *shabua* represents the most severe and universal persecution of Christians and Jews in history. As Jesus noted:

> And except those days should be shortened, there should no flesh be saved: but for the elect's sake those days shall be shortened. (Matthew 24:22)

The Archangel Heylel, failing to elevate himself above God Almighty and cast down, at this point through his minion the Antichrist, stands on the precipice of achieving his ultimate goal, the destruction of Jews and Christians to thwart the kingdom of God and Christ.

Yet for the unregenerate remaining after the catching away at the sixth seal, true wrath and destruction reflect the worst of judgments to follow during seven trumpet judgments and seven vials.

Many teachers adopt a position insisting Christians depart before "tribulation," reasoning the alternative would discourage the saints. For souls under the altar, slain for the Word of God and their testimony, Jesus did not caution them to be silent out of concern their sacrifice might discourage fellow servants and brethren facing the same death.

Masking the problem by inventing a label "tribulation saints," saved after the catching away, imposes an artificial construct unsupported in Scripture. When confronted with danger, face it head on and make the problem part of the solution. Today believers around the world face that prospect daily. If discouraging the saints represented a legitimate doctrinal backstop, why not stop talking about sin and ignore hell?

The intent here rests on clarifying a consistent sequence of events as described in Scripture. It does not hinge on a breakdown and comparison against the spectrum of teaching alternatives. Scripture regarding Christ's return stands on its own merit as to what God means when He speaks, and by the fulfillment of events themselves. As written by Isaiah,

> Seek ye out of the book of the Lord, and read: no one of these shall fail, none shall want her mate: for my mouth it hath commanded, and His spirit it hath gathered them. (Isaiah 34:16)

After signing of an agreement referred to as a "covenant with death" and unveiling of the Antichrist, the reality will be painfully obvious. Hundreds of millions of Christians around the world are already experiencing what can be referred to as severe tribulation, or the "beginning of sorrows." Yet all true believers are to be caught up before the impending wrath of the Day of the Lord, constituting events after the opening of the seventh seal (Revelation 8:1).

A final week to come can be identified with the Nation of Israel submitting to a supposed peace agreement or "covenant with death" (Isaiah 28:14-19) for the guarantee of their protection.

In Daniel 9:26, speaking of a "prince who is to come," he (the Antichrist) confirms a covenant with many for a *shabua* or seven. That may constitute ratification of an existing agreement. Scripture references the period of seven as two halves of 3-1/2 years (42 months, 1260 days) based on a 360-day Jewish prophetic calendar.

In the middle of the seven, the Antichrist breaks the covenant of peace, just as did Antiochus-IV Epiphanes with the Maccabees in the inter-testamental period. That description of the Abomination of Desolation foreshadows the future referred to by Jesus during the Olivet Discourse (Matthew 24, Mark 13, Luke 21).

Abraham's covenant with Abimelech (Genesis 21:22-33) at Beersheba (well of the oath, or seven, or completion) provides foreshadowing of the current Abraham Accords, seeking to reconcile nations, peoples, and creeds of surrounding nations to God's covenant people. Drawn by a promise of security, Israel will be led into a compromising trap as the Antichrist rises to a place of influence.

At some point in the process, the signing of such a covenant agreement by the Nation of Israel marks the commencement week or *shabua* of 2 x 1260 days (2520 days). The climax of those 2520 days represents the ultimate fulfillment of the Day of Atonement. On that day one-hundred-forty-four thousand Jewish remnant, hidden in Sela or Petra, are visited by a glorious appearing of Jesus (Revelation 14:1-5). There they come to a personal knowledge of Him as their Messiah (Zechariah 12:9-14).

Upon this sign, Christ and the remnant of Jewish believers will follow Him to confront the Antichrist and unregenerate nations. Simultaneously, in the final outworking of God's wrath, seven angels go forth to pour out seven vials or bowls of the wrath of God Almighty (Revelation 16:1-21).

Jesus Did Not Set a Date ...but a Sign

Each fall for over thirty years, I have calculated the day 2520 days before Yom Kippur seven years in the future to monitor whether a peace covenant signing takes place. There is coming a year when a covenant will be signed on that day signaling the commencement of the final seven years during which the seals will be opened.

For the year 2026, the date is November 8-9. This will be the first time in my recollection that Resurrection Day (Easter) occurs 1260 days later (April 21, 2030) and Yom Kippur (October 2-3, 2033) occurs 2560 days after the covenant date. If the "covenant of seven" confirmation occurs November 8-9, 2026, then the catching away of believers will take place during the festival of Rosh Hashanah (September 17-19, 2031).

This does not involve "date setting" but watching for a "sign" identified by an event (Matthew 24:30). Within that period of Rosh Hashanah, the Jewish idiom applies, "No one knows the hour or the day". The clock begins with a sign of signing of a peace accord as confirmation.

After the signing of that Abrahamic covenant, at the midpoint of 1260 days a desecration of sacrifice will take place. It will usher in the Abomination of Desolation and period of great tribulation. These events, revealed to Daniel and John are sealed in mystery, underpinning the plan of God for Christ's return.

Times and seasons are known, though closure of the final act remains shrouded in the purpose of God Almighty. As events begin to unfold, they execute according to strategic steps of a plan rather than disconnected chaos. Jesus challenged all therefore to;

> And what I say unto you I say unto all, Watch. (Mark 13:37).

Lunar Reflectivity

Upon posing a question to fellow scientists about lunar reflectivity, a number of technical reference papers were recommended. After some research, the study of lunar light spectrum yielded a publication and data showing the moon's spectral reflectance. The reflectance varies as a function of incoming wavelength for a range of moon surface conditions.

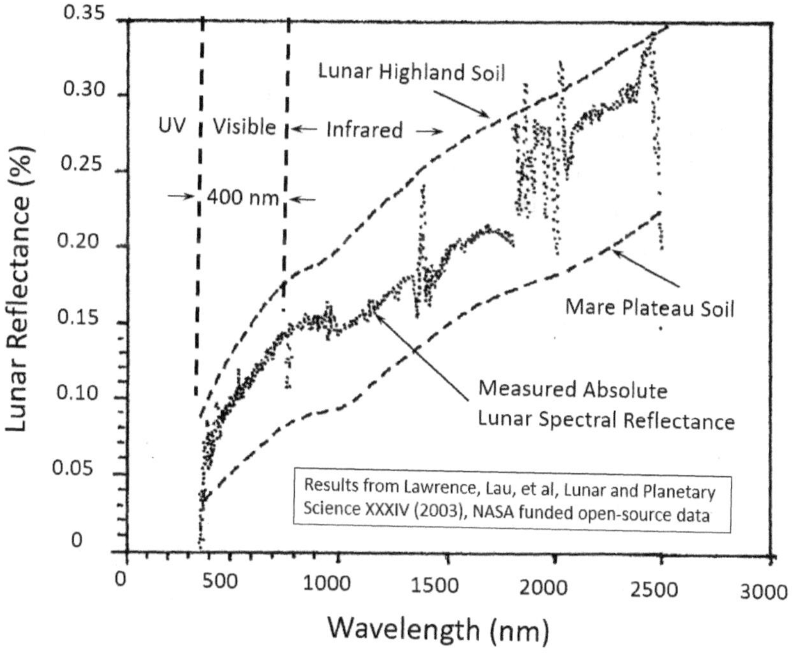

Chart of Lunar Reflectance

Results were published in "A New Measurement of the Absolute Spectral Reflectance of the Moon," S.J. Lawrence, E. Lau, et al, Hawaii Institute of Geophysics and Planetology, U. of Hawaii, Manoa, 2003.

In this paper Lawrence, Lau, et al compared the results of lunar reflectance versus wavelength for a variety of locations and also to the reflectance of lunar surface samples returned by Apollo astronauts. Measured reflectance from the moon and lunar samples showed a similar trend as a function of wavelength. In addition, results for the lunar samples provided upper and lower bound relative to reflected spectral data.

The human eye senses light wavelengths from 780 to 380 nanometers (nm = $1x10^{-9}$ m). Shorter wavelengths represent ultraviolet (UV) spectral range (UVA [400-315nm], UVB [315-280nm], UVC [280-100nm]. Longer wavelengths represent infrared (IR [780 nm to 2600 nm]). Radiation energy distribution in the solar spectral range skews toward ultraviolet and is visible as a function of the sun's equivalent radiant temperature (~5500°K).

The atmosphere reflects, scatters, and absorbs significant amounts of dangerous ultraviolet subject to environmental conditions. Solar radiation incident at the moon's surface does not experience atmospheric scattering, and hence is effectively identical to radiation in space. Overall distribution of energy from sunlight includes 5% UV, 42% Visible, and 53% IR.

Solar spectral shifting due to the heavens rending open and the resulting stretch of spacetime fabric would change light wavelengths in process from ultraviolet toward infrared. The magnitude of wavelength change would depend on the extent of the stretching.

A clue can be determined from the event recorded in the Old and New Testament prophesies. The removal of gravitational influence for distant mass resulting from rending the heavens would diminish vacuum energy density from far mass influence.

Suffice to say, the description in the above noted scriptural references of normally ultraviolet sunlight shifted to visible infrared indicates it will be consistent with a wavelength increase of about 300 nm.

Data published in the above technical paper shows lunar surface reflectivity change would be spectacular, ranging from two to four times greater reflected energy based on beginning and ending wavelengths. Though the lunar light reflected would shift to red, the amount of total energy flux reflected would increase several times based on the wavelength shift, giving rise to a blood red moon.

During the rending of the heavens, radiation including microwave, cell phone, and satellite communication will be distorted by a shift in frequency (and wavelength). Even the key fob remote and tracking devices will fail, at least for a time. As concerning the altered spectrum of light, by conservation of energy, heat transmission to the earth remains essentially the same.

When the heavens begin to open at the Lord's return, Jesus said events would unfold across the globe as lightning (Matthew 24:27, Luke 17:24). Light travels 300 million meters per second, but lightning travels about 120 thousand meters per second. Earth's circumference at the equator is about 40 million meters, which means light would circuit the globe in about one tenth of a second, but at the speed of lightning it would take almost six minutes.

The distinction becomes important considering the result. At the speed of light, no one could register what had happened. Lightning however occurs quickly, but the outcome cannot be missed. With communication disrupted by the frequency shift, there would be no prior notification around the world. But consequences of the event will result in believers, both dead and alive, being removed to prepare for the time of God's wrath; "Yom YHWH" or Day of Yahweh or Day of the Lord.

Lawrence, Lau, et al recorded relatively low lunar reflectivity for rock and dust as a function of wavelength. In this study, reflectivity measured less than 0.05 (5%) in UV to almost 0.35 (35%) in IR.

The net effect of a cosmic shift from UV to IR (~300 nm) would be dramatic with the sky turning black across the earth due to a sun darkened like reddish coals in a campfire. Consequently, the moon would become deep blood red. As the fabric of the heavens rolls away, light from heavenly bodies (including stars) moves across the sky, as Jesus said,

> For as the lightning, that lighteneth out of the one part under heaven, shineth unto the other part under heaven; so shall also the Son of man be in his day. (Luke 17:24).

Falling Stars

As noted, each star exhibits spectral characteristics of frequency distribution and intensity. During an event of rending the heavens and the resulting stretch, stars predominantly in infrared would shift to invisible long IR, visible spectrum to infrared, and ultraviolet to visible. Consequently, black skies still contain visible stars as the heavens are rolled away, giving the appearance of stars "falling."

This sequence of events unfolds across the sky and around the world. But what of places where the atmosphere is obscured by cloud cover? On an overcast day, when the sun moves low on the horizon and light passes through the atmosphere at a shallow angle, clouds frequently begin to dissipate.

This depends on atmospheric conditions, but as light at a shallow angle passes through more oxygen, UV is filtered and the remaining IR light tends to clear the clouds. In the case where all of the sunlight is shifted toward the IR range, it creates a powerful heater to dissipate overcast. Under normal atmospheric conditions, UV tends to increase humidity and convection, thus increasing cloud cover, whereas IR reduces humidity and clears the atmosphere.

Every person living during these events will experience trembling earth, blackened sky, darkening celestial lights, and rolling back of the heavens. Many will be caught away at the appearance of Jesus Christ to the sound of a great trumpet and shout of the archangel. These events quickly circuit the earth, appearing in succession to everyone in the world. Just as the heavens open, they quickly close again and nothing remains the same as God executes His final work.

Come Quickly, Lord Jesus!

Events disclosing the catching away of believers are outlined by numerous Scriptures. The Lord Jesus Christ promised to return in the heavens (Matthew 24:27 and 24:44). Paul the Apostle detailed how the Lord would descend with a shout, the voice of the archangel, and the trump of God to gather deceased and living believers (1 Thessalonians 4:16-17). In Acts 1:10, two angels declared to the disciples that Jesus would return as He left in the clouds. As quoted by Peter (Acts 2:17-21) regarding the last days, the prophet Joel prophesied,

> And it shall come to pass in the last days, saith God, I will pour out of my Spirit in those days upon all flesh: and your sons and your daughters shall prophesy, your young men shall

> see visions, and your old men shall dream dreams: And on my servants and on my handmaidens I will pour out in those days of my Spirit; and they shall prophesy: And I will shew wonders in heaven above and signs in the earth beneath; blood, and fire, and vapour of smoke. The sun shall be turned into darkness, and the moon into blood, before the great and notable day of the Lord come: And it shall come to pass, that whosoever shall call on the name of the Lord shall be saved.

Many published books and articles about the second coming of Jesus Christ argue these cosmic events in Matthew 24:29 represent Christ's physical return after the catching away to rule the earth a thousand years.

A fundamental problem with this notion rests on the virtually identical account at the opening of the sixth seal from Revelation 6:12-14, immediately followed by the Day of the Lord's wrath poured out on the earth. Clearly, signs at the opening of the sixth seal cannot be fit into any scenario associated with the Lord's triumphal return with the saints as described in Revelation 19:

> And I saw heaven opened, and behold a white horse; and he that sat upon him was called Faithful and True, and in righteousness he doth judge and make war. His eyes were as a flame of fire, and on his head were many crowns; and he had a name written, that no man knew, but he himself. And he was clothed with a vesture dipped in blood: and his name is called The Word of God. And the armies which were in heaven followed him upon white horses, clothed in fine linen, white and clean. And out of his mouth goeth a sharp sword, that with it he should smite the nations: and he shall rule them with a rod of iron: and he treadeth the winepress of the fierceness and wrath of Almighty God. And he hath on his vesture and on his thigh a name written, KING OF KINGS, and LORD OF LORDS. (Revelation 19:11-16)

Revelation 6 and Matthew 24 as well as other Old and New Testament verses describe a series of signs associated uniquely with the rending of the heavens. These signs display the return of Christ and the catching away of His people prior to the outpouring of the Day of the Lord's Wrath.

Subsequent events unfolding during the seven trumpets and the seven vials of wrath represent severe judgment upon unrepentant nations, absent believers. During this period one-hundred-forty-four thousand Jews remain sealed and sequestered in the wilderness of Moab and Edom. It is there that the remnant will be confronted by the physical return of Christ on the ultimate fulfillment of Yom Kippur.

The appearance of Jesus Christ in the heavens at the sixth seal represents a truncation before the end of the final 3-1/2 years for the sake of believers, as Jesus said,

> And except those days should be shortened, there should no flesh be saved: but for the elect's sake those days shall be shortened. (Matthew 24:22)

After the catching away, during the seven trumpets and the seven vials of wrath, a few clues appear to indicate the time frame. After the fifth trumpet (Revelation 9:1) there is a period where men are tormented by the sting of locusts or scorpions for five months (Revelation 9:5, Revelation 9:10). Also, upon the sixth trumpet, four angels are loosed for an hour, and a day and a month and a year (Revelation 9:15).

These two amount to at least eighteen months following the catching away and prior to the fulfillment of Yom Kippur as described in Revelation 14:1-5. After their acknowledgement of Jesus Christ, the one-hundred-forty-four thousand come with Christ upon Mount Sion (or Zion).

As described earlier, the completion of Rosh Hashanah and the catching away should be before this eighteen-month window during the second 3-1/2 years of the seven-year *shabua*. Therefore, the Rosh Hashanah two years prior to the final Yom Kippur should coincide with the catching away of the believers. As noted, that would be the "Day and Hour No Man Knows" because in Jewish tradition, it would be declared by the High Priest based on observation of the new moon by two witnesses. Pouring out of the seven vials or bowls of wrath then takes place after fulfillment of the final Yom Kippur (Revelation

16:1-17). These events will be part of the 75 (30+45) days preceding the celebration of Hannukah.

From Daniel 12:11-13, the Lord encouraged him concerning a hopeful event 1290 days after the Abomination and another at 1335 days. Since the final 3-1/2 years constitutes 1260 days, the remaining 30-day and 45-day periods are consistent with the 75-day period after Yom Kippur and before Hannukah at the time of Antiochus-IV.

During this time the temple was cleansed, restored, and rededicated. To this day, Hannukah always falls 75 days after Yom Kippur. So also, by completion of the Day of the Lord, all of his enemies are to be vanquished, the Kingdom of David and Solomon restored in their completion, and the final Festival of Tabernacles satisfied in rest a thousand years.

After the millennial reign of Christ occurs the final release of Satan, whereupon he will attempt to lead unrepentant nations against Christ. Satan and the nations, represented by Gog and Magog, are then destroyed by fire from heaven (Revelation 20:7-10), and afterward thrown into the lake of fire and brimstone. Next follows the Great White Throne Judgment, death and hell cast into the lake of fire, and final eternal rest in glory (Revelation 20:11-15).

Festival of Tabernacles

To encapsulate the culture of harvest, God instituted a series of seven feasts outlined in Leviticus 23, grouped into three seasons or festivals. These extended periods of harvest from spring to fall included key fruits and grains such as barley, wheat, oats, grapes, figs, pomegranates, and olives.

God called all males to gather with an offering at a chosen place (later identified as Jerusalem) three times in the year (Deuteronomy 16:16) marking harvest sabbaths of rest. Passover (Nisan 14), Unleavened Bread (Nisan 15-21), and First Fruits (Nisan 16) encompassed Passover Season, opening the harvest.

Fifty days (seven sabbaths plus a day) after Passover, the feast of Pentecost (Sivan 6) or shavout commenced, marked by a wave offering of two loaves of bread. These symbolized the entry of Jews and Gentiles into faith. Pentecost memorialized receipt of the table of law by Moses, looking forward to the law written in the heart by the Holy Spirit.

A season of Tabernacles encompassed the longest period of assembly beginning with Trumpets (Tishri 1) or Rosh Hashanah (meaning shout or alarm). Next followed the Day of Atonement (Tishri 10) or Yom Kippur (meaning day of expiation). The final feast of the season was Tabernacles itself (Tishri 15-21) or Booths (meaning tabernacle).

From the first day of the seventh month (Tishri, first civil month), to the final sabbath (Leviticus 23:39) on the twenty-second day of Tabernacles represented closure of the latter harvest. This eighth day sabbath, spoken of by the Lord, alludes to further rest after the millennium.

The feast of Tabernacles constitutes the only symbolic festival carried out during the millennial reign of Christ (Zechariah 14:16), which represents a looking forward to final rest in God's presence in the heavenly realm.

In addition to the seven Feasts of the Lord, Scripture and Jewish tradition include others such as Tisha B'Av (Av 9), Hanukkah (Kislev 25 – Tevet 2/3), and Purim (Adar 14). Jewish festivals instituted by the Lord in Leviticus 23 reflect the call of Jesus in John 4:

> Say not ye, There are yet four months, and then cometh harvest? Behold, I say unto you, Lift up your eyes, and look on the fields; for they are white already to harvest. And he that reapeth receiveth wages, and gathereth fruit unto life eternal: that both he that soweth and he that reapeth may rejoice together. And herein is that saying true, One soweth, and another reapeth. I sent you to reap that whereon ye bestowed no labour: other men laboured, and ye are entered into their labours. (John 4:35-38)

Return to Normal?

After the catching away of believers, what next? Does the cosmos and the world return to some semblance of normalcy? Just as the heavens rolled back to herald the appearance of Jesus Christ, the cosmic fabric again closes. At this, light from the sun, moon, and stars return to their former state, but nothing else resembles normal. After unprecedented events during the sixth seal, all people remaining cry out:

> ...Fall on us, and hide us from the face of him that sitteth on the throne, and from the wrath of the Lamb: for the great day of his wrath is come; and who shall be able to stand? (Revelation 6:16,17)

Suddenly multitudes clamoring against God and the faithful in Christ will get "what they demanded". With true and faithful believers gone, the world senses foreboding and a harbinger of destruction to come. Beginning with the seventh seal, the ensuing seven trumpets, and the seven vials pour out the wrath of God upon the disobedient and rebellious.

Attempts at Israel-Palestine peace agreements the past fifty years include Camp David Accords (1978), Madrid Peace Conference (1991), Oslo Accords (1993, 1995), Camp David Summit (2000), and Abraham Accords (2020). Negotiations in 2025 and 2026 seek to expand participants to the Abraham Accords in hopes of a settlement to Israel-Arab conflict. The current Abraham Accords, patterned after Abraham's covenant with Abimelech at Beersheba (well of the oath, seven, or completion), lay the framework for the final covenant of seven spoken of by Gabriel (Daniel 9:27).

At some point in the near future, this framework covenant will serve as basis to constrain Israel under an agreement to be confirmed by Antichrist. As prophesied by Isaiah, this will be a "covenant with death" or "agreement with hell" (Isaiah 28:15-18). The Lord reveals his work in an appointed time, though clearly the time is short.

After revealing of the Antichrist figure, and confirming a covenant with Israel, those who misinterpreted the message will require mental gymnastics trying to process how they missed the obvious. Jesus will appear suddenly at a time appointed by God Himself. As in the parable of the virgins, "...at midnight there was a cry made, Behold, the bridegroom cometh; go ye out to meet him." (Matthew 25:6).

Time to get ready. Even so come, Lord Jesus.

Greek Amphitheater and the Play *L'Acedia, Land of Leisure*

Chapter Eighteen - Rend the Veil

Scarcely had there been time to clear rubble and debris from the streets. Portions of the City of L'Acedia, lay in ruin. Rioting and looting followed the latest outbreak of conflict, yet a movement already began to re-establish order in L'Acedia.

As Act III of the play *L'Acedia, Land of Leisure* continued after intermission, a few members of the audience shuffled back to their seats. Fleeting rays of sunshine hung on the recently restored outdoor amphitheater. The final act opened to a few conspicuous changes onstage; others manifested out of sight behind a large veil.

Evangels and Bread Distribution Tables

Act III, Scene 1: Calm after the Storm

In the midst of the cleanup, a small group of church-goers assembled with tables offering an abundance of food, drink, and necessities. The tables provided a lifeline to citizens of L'Acedia after most of the stores closed during recent rioting. Townspeople labelled them "Evangels" for their incessant portrayal as so-called "good messengers". Their unregulated table distributions created a new dispute.

> One of the citizens in the breadline commented, "These Evangels take a big risk trying to give out food and drink without a permit." Another responded, "Regardless, I am grateful they have access to bread with food now tightly controlled."

The City of L'Acedia After Rioting

No sooner had a line formed at one of the distribution tables but a much larger collection of citizens and outsiders began to gather. Some of the developing crowd appeared recognizable, despite black hoodies and facial masks. Many had participated in the unrest, even cheering rioters on at times.

Hanging a Banner Next to the Green Fairy Salon

Quickly, a few outsiders calling themselves the Queen of Hearts Posse, unfurled a banner beside an emerald-colored storefront of the Green Fairy Salon. The Green Fairy Salon (or GFS), had become a long-standing lair for poets, playwrights, and the disenfranchised of L'Acedia.

Written on the new banner was a slogan with a heart-shaped serpent logo, and wording vaguely reminiscent of the French Revolutionary motto.

ASP – Anarchy, Sodality, Parity ...*or death*

The Green Fairy Salon

Work of Darkness

Their black hoodies emblazoned with the Queen of Hearts Posse emblem now contained a subtitle, "Off with Their Heads"; a not-so-subtle reference to recent calls for violence against anyone opposing the uprising.

"It appears '*Mr. Work of Darkness*' now funds these anarchists", whispered someone else in the breadline, alluding to the inference in the name Ergon Schwartz.

The Queen of Hearts Posse and Ergon Schwartz

Rumors persisted that Ergon Schwartz, the shadowy international financier, supported the outside anarchists of the Queen of Hearts Posse. Ergon and his son Aleksander now controlled much of the commerce in L'Acedia as a result of their business called Association for Social Progress (ASP).

Ergon managed to undermine and topple leadership of business and government within L'Acedia, building a fortune in the process. Using the classic "short and distort" technique of underbidding the value of a business, and then planting subversive rumors, Ergon and his son Alek managed to buy up most of their competitors. Each of Ergon's financial endeavors, whether business or benevolence enabled him to gain a stranglehold on the affairs of the aptly named LA-LA Land.

ASP Evolves into GRASP

Recently, the father-son pair of financiers set their sights on world markets. Ergon Schwartz and his son Alek restructured their organization into an international network, Global Renewal Association for Social Progress (GRASP). The growing influence of GRASP appeared to be reminiscent of Mr. Potter's hold on Bedford Falls in the movie, *It's a Wonderful Life* (1946). Aleksander Schwartz maintained ambitions of worldwide influence, as did his namesake, Greek conqueror Alexander the Great.

Edward Alexander Discards the Script

To the side of the stage, next to an empty director's chair sat a trash container. A member of the radical old guard, Edward Alexander walked briskly forward, sat in the director's chair and threw his copy of "The Script" into the wastebasket,

> "No use for this outdated script anymore!", Edward boasted. "It appears the Director dis-appeared, and we have rewritten the script!", he continued.

Edward Alexander, raised by strict Evangel parents, rebelled from an early age. Rejecting his Christian upbringing, Alexander instead patterned his ambition after his name as, *Defender of Man*. He proudly embraced his mother's label of disdain for her son's behavior as "The Beast".

Wormwood

Known internationally as explorer, writer, and occultist, Edward Alexander frequented The Green Fairy Salon for absinthe, his drink of choice and namesake for the establishment. Alexander's sinister reputation earned him disdain and the title, "The Wickedest Man Alive."

His support for Ergon Schwartz opened the door for Edward to gain a foothold in L'Acedia. Public opinion often attributed Edward Alexander's erratic behavior to his fondness for absinthe, also labeled the "emerald muse" or "wormwood".

One of Edward Alexander's controversial ideas involved a "new religious" belief system he labeled "Thema". His mystical ideas around finding and doing one's own will developed a cult-like collection of followers in L'Acedia.

Absinthe "Green Fairy" Liqueur

Bottled absinthe, by nature bitter and of very high alcohol content, is traditionally diluted with chilled water drizzled over a sugar cube on an absinthe spoon. Absinthe received its name and dark reputation due to brewing of alcohol with herbs of wormwood (artemisia absinthium), anise, and fennel, yielding an emerald green spirit. When served, emerald green absinthe turns into a cloudy white liqueur.

Often claimed to cause hallucinations, absinthe received the moniker, "the green fairy". For that reason, much of Europe banned absinthe during portions of the nineteenth and twentieth centuries until processing and content were controlled. With its reputation for enhancing creativity, absinthe gained a loyal following among L'Acedia's poets and artists of the Bohemian fringe.

Max Opens the Green Fairy Salon

Act III, Scene 2: The Green Fairy Salon

By mid-morning, Max, the proprietor, prepared to open GFS for a small gathering of early customers. Meanwhile, Aglaope the greeter seated herself strategically near the doorway. Renowned for her engaging appearance and melodious siren's voice, Aglaope (Uh-glay'-oh-pee) consistently charmed wary passersby, converting them into loyal clients. As one of the few establishments with a reliable source of food and drink, GFS attracted most of the local populace as well as patrons from every district and culture.

Aglaope the Greeter Seated Outside the Green Fairy Salon

Over time the Green Fairy Salon earned a reputation as the locus of unrest in L'Acedia. Regular gatherings in the salon over the past year led by Edward Alexander fueled demands for action against the few remaining but influential churches and synagogues. After recent violence, a few salon patrons had been accused of participating in covert acts of aggression against places of worship. Though not directly involved in the unrest, Max became a passive enabler.

Patrons Seated Outside the Green Fairy Salon

By noon patrons filled the outdoor seating at GFS. Increasingly, black-hooded members of the Queen of Hearts Posse were making GFS their gathering place. With most establishments in L'Acedia closed, the Green Fairy Salon continued to gain a loyal following.

The Kandi Kollective Representatives

Act III, Scene 3: The Agreement

Suddenly there appeared three whimsically attired strangers. The newcomers, Wink, Blink, and Nod introduced themselves to Max, and proposed a confidential business negotiation.

> Their spokesman, Nod opened, “We represent the Kandi Kollective.” Blink added, “Your bargaining contract for food, drinks, and labor will expire shortly.” Wink continued, “In order to keep your salon operating in the face of a challenging social climate, we have a helpful proposition that will ‘candy-up’ your establishment’s standing.”

The Sucker Guild

The Kandi Kollective, often labeled the “sucker guild”, began by representing bargaining interests of confectionary and sugar production workers. Recently the Kollective muscled into other avenues of commerce, eventually obtaining controlling influence over most of the food and beverage industry in the province surrounding L'Acedia. Behind the scenes, Ergon Schwartz bankrolled most of the initiatives of the Kandi Kollective, enabling him to also influence the labor market.

It appeared continuing violence in L'Acedia had restricted distribution of food and beverage shipments to the point where Kollective workers were impacted. Evangel members and supporters, despite efforts to assist the city's populace, now created competition to remaining business and labor initiatives. The trio representing the Kandi Kollective offered Max a lucrative opportunity to expand influence throughout L'Acedia and squeeze out the distribution tables.

Oskar "The Swindler" Rag Man and His Pull Cart

Oskar Swindler

In the middle of the discussion, a handcart moved slowly down the street. Piled with old cotton rags, Oskar "Swindler" pulled the cart while dressed in a vintage suit and black top hat.

Oskar appealed, "Rags! Rags! Old cotton rags!"

Though appearing to be of meager means, Oskar attracted criticism by buying old cotton clothing for next to nothing, bundling them, and selling bales of cotton rag for reprocessing. With the demand for high quality cellulose material in the manufacture of smokeless gunpowder (nitrocellulose), Oskar managed to make a small fortune out of old cotton clothing that many discarded as trash. Oskar the rag man's unassuming appearance managed to avoid political scrutiny as well as the clutches of Schwartz and GRASP.

To his credit, Oskar created work for a select few workers and devoted a portion of his earnings to support elderly widows in the city. Some of the agitators in L'Acedia referred to them as hags. A few envious citizens of L'Acedia gave him the name "the swindler", and it stuck. Occasionally citizens and outsiders chanted at the passing handcart.

"Oskar Swindler! Cheat the poor and support the war!
Oskar Swindler! Sell your rags to help the hags."

Max and The Kandi Kollective Proposal

Ignoring the interruption, negotiations continued. The trio expanded on the terms of their offer to Max and The Green Fairy Salon.

> Nod assured Max, "The Kandi Kollective will arrange to supply food and drink to GFS at an exclusive discount well below wholesale prices, provided you use the salon to implement a community-wide membership agreement with patrons."

Max listened attentively until this point, but the thought of convincing his libertarian customer base to accept any kind of agreement seemed a daunting challenge.

> Voicing his concern, Max argued, "This crowd of open-minded thinkers are more difficult to corral than a flock of free-range chickens. I don't see any way to create an agreement that will not alienate the majority."

To that objection, Wink, Blink, and Nod were swift to offer a strategy.

Wink proposed a subversive step, "First, recruit several of your most loyal patrons to infiltrate the competition and gain their confidence, while asserting your influence, and sowing discord."

Blink then offered a diversion tactic, "Next, start a campaign of rumors and contaminate the competition's source of supply."

Nod delivered the concluding blow, "Finally, implement an emergency rule with the help of local politicians requiring everyone to register and obtain a number to buy or sell. Those who do not comply will be subject to confiscation of goods, fines, and imprisonment with required behavioral retraining!"

At that point Max showed interest, but objected, "I am not in good standing with L'Acedia's leaders. How can I get them to work with me?"

Wink, Blink, and Nod spoke in succession. Wink explained, "That's where we come in." Blink offered, "The Kandi Kollective has the local politicians in our pocket." And Nod added, "We guarantee their cooperation."

Max continued to stall, protesting, "Give me a week to think about your offer. I will discuss it with my staff and patrons."

Nod stated flatly, "We are no longer speaking about an offer. It would be a shame to see the Green Fairy Salon go out of business …or suffer the same fate as your neighbors."

With no other options, Max nodded to the affirmative.

Nod displayed their standard contract, stating simply, "So it's agreed then."

Act III, Scene 4: Return to Normal

With the cleanup effort over the ensuing weeks, streets of L'Acedia gradually returned to some semblance of normal, though the atmosphere darkened noticeably. Distribution tables had all but disappeared. Many of the Evangels abandoned the tables, troubled by unrest within their ranks. Accusations of contaminated bread and drink drove away those in need.

The Green Fairy Salon, however, enjoyed brisk activity. Forced registration and lack of competition brought a steady stream of new patrons, despite the higher prices. However, success created a dilemma of sorts. A growing number of citizens who had been turned away because they refused to register, were now causing vocal disturbances. Most of the holdouts were Evangels and Jews.

The Green Fairy Salon Registration Table

Adding to the social tension in L'Acedia, an abandoned place of worship suddenly began an extensive renovation project. Rumors circulated that the secretive financier, Ergon Schwartz's organization funded a complete remodeling of the structure into an exclusive sporting complex and spa.

In conjunction with his son Aleksander Schwartz assuming greater control of the international expansion or GRASP, the fitness center would be aptly named Aleksanders Klassik Athletiks (or AKA). Alek planned to revive the classic Greek gymnasium, promoting AKA sports complex as an Olympian clothing optional center.

Sculpture Model Display at GFS

With work on AKA now approaching completion, a local artist and patron of GFS announced he had been commissioned to create a sculpture adorning the building's atrium. Some of L'Acedia's residents expressed displeasure at the advertisements announcing the AKA complex as "Clothing Optional". Other citizens dismissed the notice as only an advertising ploy. Either way, it succeeded in creating a great deal of interest and controversy

Max agreed to display a scale model of the sculpture in front of The Green Fairy Salon to build enthusiasm. Not surprisingly, the facial appearance of Ergon Schwartz adorned the life-sized image of Zeus. As the grand opening of AKA neared, local leaders scheduled an unveiling in the theater on Easter, coinciding with completion of citizen registration.

Anti-Registration Re-education Asylum

The Transformation Asylum

To address the problem of unrest over the registration law, a hastily enacted ordinance by the leaders of L'Acedia mandated rehabilitation for all citizens without a registration number. Law enforcement personnel repurposed an abandoned school facility for the social engineering project

Ergon Schwartz, in his usual subversive manner bribed local officials to allow members of the Queen of Hearts Posse to fill most of the positions. Concerned family members occasionally gathered at the entrance gate of the facility only to be turned away by members of the Queen of Hearts Posse.

It became evident Schwartz and his son had obtained control of both the registration and re-education processes. As days extended into weeks and then to months, rumors in L'Acedia questioned why no one appeared to graduate from the rehabilitation school.

Despite escalating concerns about the rehabilitation facility, Oskar the rag man came forward with an attractive proposition. Oskar's business already began to suffer from a lack of qualified personnel.

In order to maintain his contracts for supply of cotton rags, Oskar approached one of the local leaders with a proposal.

> "I must have additional workers to sort rags or risk missing my contract deliveries." Oskar continued, "I will make it well worth your while if you allow me to choose several of those being sent for rehabilitation."
>
> Without hesitation the politician agreed, arrangements were made, and "the piper was paid".

Quickly Oskar began escorting away selected candidates before they were to be confined for re-education. Without any fanfare, Oskar managed to identify and remove a number of Jewish citizens destined for rehabilitation.

Act III, Scene 5: Setting the Stage for the Unveiling

Tensions in L'Acedia and the surrounding Land of Leisure finally came to a boiling point. Many citizens agonized over the continuing failure of the Re-education Asylum to release those who were sentenced for not registering. A dwindling food supply and the lack of activity at the asylum fueled rumors that the detainees had simply "disappeared".

By now Ergon Schwartz effectively controlled all of L'Acedia through a series of carefully orchestrated political and financial maneuvers. The man who labored to unravel the political, social, and religious fabric of L'Acedia, had transformed it into a dystopian nightmare rather than a promised social paradise.

Dwindling numbers of non-registered citizens created a chilling awareness throughout L'Acedia despite regular reassurances from local leaders. The office of Ergon and Aleksander Schwartz did little to quell the uneasiness. Even media outlets manipulated their messaging to minimize any criticism of GRASP and the agenda of Ergon and Alek Schwartz.

With the opening of AKA Fitness and Spa approaching at Easter, preparations continued to put the finishing touches on the atrium including installation of the imposing sculpture. Though still veiled in a shroud, the statue's shape appeared oddly unlike the model displayed in front of the Green Fairy Salon.

The Curtain is Torn

Act III, Scene 6: Deus Ex Machina on a Cosmic Scale

As the crowds returned to their seats for the final scene at L'Acedia theater, excitement grew in anticipation of the statue unveiling. Suddenly the ground began to shake as Ergon and Alek Schwartz entered the stage for the introduction ceremony. A collective gasp and shouts arose from the audience as the earth quaked.

To the amazement of viewers, the veil backdrop at the rear of the stage tore from top to bottom revealing an empty room. A rumble of whispers filled the theater while a small curtain at the left side of the stage opened slowly to reveal the Jester at the theater controls. With the pull of a lever, the imposing image of a statue covered with a shroud slid to the forefront.

The Jester Behind the Curtain

An attendant pulled back the shroud revealing the image; but not the highly anticipated sculpture of Zeus nor the likeness of Ergon Schwartz. The figure appeared to be a representation of the Jester himself. From the audience arose a chorus of disapproval as Ergon and Alek Schwartz fled the stage.

A Scandalous Image

Always quick to counter insults, the Jester launched into an impassioned routine of profanity laced humor, loaded with sarcasm. His demeanor and delivery caught the attendees off guard. Uneasy laughter punctuated the audience. Just as quickly as his monologue began the ground again trembled and the skies darkened.

Cosmic Deus Ex Machina

To everyone's amazement, the heavens quickly parted while the sun dimmed and the moon turned a deep blood red against an ebony sky. Stars appeared to fall across a blackened scene moving across the heavens. Quickly the realization set in throughout L'Acedia; this was the commencement of the Day of the Lord, spoken of by the Evangels for centuries.

> ...Surely I come quickly. Amen. Even so, come, Lord Jesus. The grace of our Lord Jesus Christ be with you all. Amen. (Revelation 22:20, 21)

The Bus to Babylon

Epilogue

Like an intoxicated bus driver on the wrong side of the road, the world system continues to accelerate out of control. Babylon, biblical symbol of a fallen world, unknowingly finds itself on a collision course with reality. Not surprising, most passengers goad the driver on, forcing opposing traffic off the road.

By contrast, the Kingdom of God portrays a scene acted out on a grand stage, conceived by a creative Author, assembled by a Master Builder, populated by a vast array of costumed players, each with particular vulnerabilities. Action takes place on the stage populated like a chess board adorned with a troupe of pieces.

As the story plays out and the plot develops, surrounded by intrigue enters the pretentious Jester Heylel, bound by a singular purpose: to promote resistance, and in process of time, distract the attention of all present from the underlying plan.

American Revolution Chess Game

Characters weave in and out of the story as events unfold, all taunted by the Jester; a few departing the scene celebrated. Most depart unceremoniously, reciting lines but never understanding the purpose. Sadly, many participants neglected to read and interpret the script; others informed, misunderstood the meaning.

Suddenly, on the cusp of chaotic rebellion, as the Jester attempts to lay hold on the crown, curtains come down and celestial lights dim. Distracted, players exit the stage as the realization settles in; it is not a board game, international intrigue, nor theater, but a reset of history as we know it.

The ultimate Deus ex machina on a cosmic scale; an improbable resolution to an impossible situation as the Author Himself briefly reenters the stage to foil the Joker and complete the supreme purpose: redemption and judgment according to His eternal plan.

REST
AREA
EXIT HERE

Rend the Veil

…the Stage is Set

www.ingramcontent.com/pod-product-compliance
Lightning Source LLC
LaVergne TN
LVHW010712061126
851857LV00014B/1156